# INTERNATIONAL ENCYCLOPAEDIA
# OF
# AGRICULTURAL GEOGRAPHY

# International Encyclopaedia of Agricultural Geography

MANISH DUBEY

CYBER TECH PUBLICATIONS
4264/3, Ansari Road, Daryaganj, New Delhi-110002 (India)
Ph.: 011-23244078. Fax: 011-23280028
E-Mail: cyberpublicationsdelhi@yahoo.com

*International Encyclopaedia of Agricultural Geography*



First Edition 2009

Published by G.S. Rawat
**For Cyber Tech Publications**
4264/3, Ansari Road, Daryaganj, New Delhi-110002
Ph.: 011-23244078. Fax: 011-23280028
E-Mail: cyberpublicationsdelhi@yahoo.com

**Printed at**
Asian Offset Printers

# Preface

Around ten to twelve thousand years ago, human began to domesticate plants and animals for food. Before this first agricultural revolution, people relied on hunting and gathering to obtain food supplies. While there are still groups of hunters and gatherers in the world, most societies have switched to agriculture. The beginnings of agriculture did not just occur in one place but appeared almost simultaneously around the world, possibly through trail and error with different plants and animals or by long term experimentation.

Between the first agricultural revolution thousands of years ago and the 17th century, agriculture remained pretty much the same. In the seventeenth century, a second agricultural revolution took place which increased efficiency of production as well as distribution which allowed more people to move to the cities as the industrial revolution got under way.

The eighteenth century's European colonies became sources of raw agricultural and mineral products for the industrializing nations. Now, many of the countries which were once colonies of Europe, especially those in Central America, are still heavily involved in the same types of agricultural production as they were hundreds of years ago. Farming in the twentieth century has become highly technological in more developed nations with geographical technologies like GIS, GPS, and remote sensing while less developed nations continue with practices which are similar to those developed after the first agricultural revolution, thousands of years ago.

About 45% of the world's population makes their living through agriculture. The proportion of the population involved in agriculture ranges from about 2% in the United States to about 80% in some parts of Asia and Africa. There are two types of agriculture, subsistence and commercial.

There are millions of subsistence farmers in the world, those who produce only enough crops to feed their families. Many subsistence farmers use the slash and burn or swidden agricultural method. Swidden is a technique used by about 150 to 200 million people, and is especially prevalent in Africa, Latin America, and Southeast Asia. A portion of land is cleared and burned to provide at least one and up to three years of good crops for that portion of land. Once the land can no longer be utilized, a new patch of ground is slashed and burnt for another round of crops. Swidden is not a neat or well-organized method of agricultural production by it is effective for farmers who don't know much about irrigation, soil, and fertilization.

Our agricultural system is in a transitional stage. Traditional agriculturists are changing some practices and their attitudes. Alternative agricultures have appeared. This book for several audiences in the agricultural and nonagricultural communities. The educators, extension workers, farmers, food and fiber processors, researchers and students in the agricultural community, and those beyond who share an interest in and a concern for agriculture.

— *Editor*

# Contents

# 1

# Introduction

Agriculture is the main stay of India's economy. Nearly two-thirds of its population depends directly on agriculture for its livelihood. It accounts for 26 per cent of the gross domestic product. It ensures food security for the country and produces several raw materials for industries. Agricultural development, therefore, is a precondition of our national prosperity.

## Land under Agriculture

The term agriculture is derived from two Latin words, *ager* meaning 'land' and *cultur* meaning 'cultivation'. Agriculture, thus, means land cultivation. However, agriculture includes animal husbandry, forestry and pisiculture. The land cultivated in a year is called net sown area. The net sown area and the land cultivated more than once together make gross cultivated area. The net sown area in India today is about 143 million hectares. It is about 46.6 per cent of the geographical area. About 23 million hectares of land remains fallow, which is about 7.1 per cent of the reported area. It varies between 5 and 7 per cent from year to year. Although India is the seventh largest country in terms of area, it tops in the world in percentage of net sown area.

The proportion of net sown area to geographical area varies greatly from 3.4 per cent in Arunachal Pradesh to 84.2 per cent in Punjab. The alluvial tracts of northern India, extending from Punjab to West Bengal and Assam and the coastal plains from Gujarat in the west to Orissa in the east are intensely cultivated.

From agricultural point of view, India is a unique country. It has vast expanse of level land, rich soils, wide climatic variations suited for various types of crops, ample sunshine and a long growing season. Agriculture has advanced considerably after independence and its contribution to export trade of India is also significant.

## Main Features of Indian Agriculture

In spite of the fact that large areas in India, after independence, have been brought under irrigation, only two-fifths of the cropped area is actually irrigated. Use of fertilizers and pesticides has increased and large areas have been brought under high yielding variety of seeds. These led to green revolution in several parts of India. This has considerably helped in increasing yields per hectare as well as total production of different crops. In spite of these developments, the country has still subsistence agriculture in certain parts. Farmers own small pieces of land and grow crops primarily for their own consumption. Animals play a significant role in various kinds of agricultural operations. Farming depends mainly upon monsoon rain. Most of the production comprises food crops. About one-third of the land holdings are small, less than one hectare in size. Means of transport in villages are inadequate and the roads are unmetalled. Even storage facilities for crops are inadequate.

## Types of Farming

Based primarily on nature of land, climatic characteristics and available irrigational facilities, the farmers in India practice different types of farming.

*Substance Farming*: Majority of farmers in the country practice subsistence farming. It is characterized by small and scattered land holdings and use of primitive tools. As the farmers are poor, they do not use fertilizers and high yielding variety of seeds in their fields to the extent they should do. Facilities like electricity and irrigation are generally not available to them. These result into low productivity. Most of the food production is consumed by the farmers and their families. Where facilities like irrigation and electricity are available, farming has improved. Important cash crops like sugarcane, oilseeds, cotton and jute are grown. The subsistence agriculture has given way to commercial agriculture to some extent.

Dry land farming is practiced in areas where the rainfall is low and irrigation facilities are inadequate. Here, emphasis is laid on conservation of moisture, and on crops like jowar, bajra and pulses, which need less water. Wetland farming is practised in high rainfall and irrigated areas. Rice, sugarcane and vegetables are important crops in these areas. In dry farming, only one crop is grown while in wet farming, at least two crops are raised in a year – one in the kharif and another in the rabi seasons.

*Shifting Agriculture*: In this type of agriculture, first of all a piece of forest land is cleared by felling trees and burning of trunks and branches. After the land is cleared, crops are grown for two to three years and then the land is abandoned as the fertility of the soil decreases. The farmers then move to new areas and the process is repeated. Dry paddy, maize, millets and vegetables are the crops commonly grown in this type of farming. The per hectare yield is low.

*Plantation Agriculture*: Plantation farming is bush or tree farming. It was introduced by the British in the nineteenth century. It is a single crop farming of rubber, tea, coffee, cocoa, spices, coconut and fruit crops like apples, grapes, oranges, etc. It is capital-intensive and demands good managerial ability, technical know-how, sophisticated machinery, fertilizers, irrigation, and transport facilities. Some of the plantations like tea, coffee and rubber have a processing factory within the farm itself or closer to it. This type of agriculture has developed in hilly areas of north-earth-eastern India, sub-Himalayan West Bengal and in Nilgiri, Anamalai and Cardamom hills in peninsular India.

*Intensive Farming*: In areas where irrigation has been possible, the farmers use fertilizers and pesticides on large scale. They have also brought their land under high yielding variety of seeds. They have mechanized agriculture by introducing machines in various processes of farming. These have led to intensive farming where the yield per unit areas is high. In some areas, this has led to the development of dairy farming.

## Major Crops

In India, there are three crop seasons, namely kharif, rabi and zaid. Kharif season starts with the onset of the monsoon and continues till the beginning of winter (June-July to October-November). Crops such as rice, maize, millets, cotton, groundnut, moong, urad, etc., are grown in this season. Rabi season starts with the beginning of winter and continues till the beginning of summer (October-November to March-April). Crops such as wheat, barley, gram and oilseeds (linseed, rape and mustard) are grown in this season. Zaid is a short crop reason of summer. Watermelons and cucumbers are grown on the sandy tracts along the river channel. Rice, maize, vegetables and fruits are major crops.

India produces a variety of crops which may be grouped conveniently under different heads, namely:

- Cereals
- Pulses and oilseeds
- Fibre crops
- Beverage crops
- Cash crops

## Cereals

Cereals cover about three-fourths of the total cropped area of the country and contribute to about half the total volume of agricultural production. Important crops are rice, wheat, millets, and maize.

*Rice*: India is the second largest producer of rice in the world after China. It is grown on about one-fourth of the total cropped area and provides food to about half of the country's population. Rice is a tropical plant. It requires high temperature and high humidity for its successful growth. The mean monthly temperature of about 24°C. with minor variations during the sowing, growing and harvesting season, is ideal for its cultivation. It grows well in areas of 100 cm of rain. In areas of less rainfall, particularly in Punjab. Haryana and western Uttar Pradesh, it is grown with the help of irrigation.

Rice is very widely grown in the country. The major rice producing states are West Bengal. Uttar Pradesh, Andhra Pradesh, Punjab and Tamil Nadu. Most of the production is locally consumed; very little is left for export. It is exported to South-East Asian countries, United States and United Arab Republic.

*Wheat*: It is the second most important food crop of India, particularly of the people of north and north-western part of the country. India is the second largest producer of wheat in the world, next to China. Wheat is a rabi in the world, next to china. What is a rabi crop in India. The ideal temperature at the time of sowing is 10 to 15°C and at the time of harvest 20 to 25°C. The crop grows well in areas where rainfall is about 75 cm: 100 cm of rain is the upper limit. It requires well-drained fertile soils. It demands comparatively less labour in comparison to rice cultivation. It is grown on about one-sixth of the cropped area of the country.

Wheat cultivation is mainly confined to the north-western part of India. The major wheat producing states are Uttar Pradesh, Punjab. Harynaa. Rajasthan and Bihar. The country is now in a position to export wheat.

*Millets*: Jowar, Bajra and ragi are some of the important millets grown in India. Jowar is the third most important food crop both in respect of area and production. It grown in areas with a mean monthly temperature of 18 to 32°C. It requires about 30 to 60 cm of rain. Major producers are Maharashtra. Karnataka. Andhra Pradesh, Tamil Nadu and Madhya Pradesh.

Bajra is also a crop of dry ad warm climate. Climatic conditions are almost similar to that required for jowar. Rajasthan is the largest producer of bajra. Other bajra producing states are Uttar Pradesh Maharashtra. Gujarat and Haryana.

Ragi is a rainfed kharif crop, grown on well drained alluvial loams, and red or black sandy loams. It requires climatic conditions like bajra and I s grown mainly in drier parts of Karnataka and Tamil Nadu.

*Maize*: It is a coarse grain, used both as food and fodder. It is grown under varied climatic and soil conditions. It is cultivated mainly as a kharif crop. It grows well in areas of 50-100 cm of rain and, in areas of less rain, it is grown under irrigation. It grows well under temperature between 21 and 27°C. It requires well-drained fertile soil. It is grown on about four per cent of the cropped areas.

Major maize producing states are Karnataka. Uttar Pradesh, Bihar, Andhra Pradesh and Madhya Pradesh.

## Pulses and Oilseeds

*Pulses*: These are the major source of protein in a vegetarian diet. Major pulses are *tur* or arhar, urad, moong, masur, peas and gram. The first three are grown as kharif crops and the last three as rabi crops. These are grown all over the country, except in areas of heavy rain. Plants of pulses help in restoring fertility of the soil and, hence, they are grown in rotation with other crops.

Major gram producing states are Madhya Pradesh. Uttar Pradesh and Rajasthan. Other important producers are Maharashtra an Karnataka. *Tur* is produced primarily in Maharashtra and Karnataka. *Tur* is produced primarily in Maharashtra. Uttar Pradesh, Karnataka, Madhya Pradesh and Andhra Pradesh.

*Oilseeds*: Vegetable oil is an important item of Indian food. Oilcake, the residue after oil is extracted from the seeds, is an important cattle feed. India is the largest producer of oilseeds in the world. Major oilseeds are groundnut, sesamum, rapeseed, mustard, linseed and castor seed. Together, these cover about 12 percent of the net sown area. Other important oilseeds are soyabean, cotton seed, sunflower and saf-flower.

Groundnut is the most important oilseed of India. It accounts for about half of the major oilseeds produced in the country. Its major producers are Andhra Pradesh. Tamil Nadu, Karnataka, Gujarat and Maharashtra. These states account for about 85 per cent of the national production.

Sesamum is a rainfed crop. It is grown as a kharif crop in the north and as a rabi crop in the south. It is grown in almost all parts of India, but Gujarat, Tamil Nadu, West Bengal, Karnataka, and Madhya Pradesh are the major producers.

Rapeseed and mustard are two most important oilseeds of northern India. These are grown mainly as rabi crops. Sometimes they are also raised as mixed crops. Major producers are Rajasthan, Uttar Pradesh, Haryana, West Bengal and Madhya Pradesh.

Linseed is grown as a rabi crop. India is the third largest producer of linseed in the world after Russia and Canada, Madhya Pradesh. Uttar Pradesh, Bihar, Chhattisgarh and Maharashtra are the major producers.

Castor seed is a small tree crop. It is grown both as kharif and rabi crops. By producing one-fifth of the world total, India ranks second in the world after Brazil. Gujarat, Andhra Pradesh and Rajasthan are the main producers.

### Fibre Crops

Cotton and jute are two important fibre crops of India. Cotton is grown in tropical and subtropical areas. It requires at least 210 frost free days in a year. It is a kharif crop, and requires 6 to 8 months to mature. At the time of harvest, the weather should be fine to maintain luster of the cotton. It grows well in black and alluvial soils. It requires a lot of cheap and efficient labour at the time of plucking of cotton balls. It is grown mainly under irrigation.

India is the third largest producer of cotton, after China and the Unites States of America. The main cotton producing states are Maharashtra, Andhra Pradesh, Haryana, Punjab and Gujarat.

Jute is a crop of the hot and humid climate. The crop takes 8 to 10 months to mature. The fibre is obtained from the retting of the plant. Almost all jute of India is produced in five states of West Bengal, Bihar, Assam, Orissa and Meghalaya.

### Beverage Crops

Tea and coffee are the important beverage crops. Tea cultivation in India is a classic example of plantation agriculture. Tea is processed from the tender leaves of a bush. As a tropical as well as sub-tropical plant, tea thrives well in hot and humid climate. Ideal temperature for its growth is 20 to 30°C and an annual rainfall of 150 to 300 cm. High humidity is good for the rapid development of tender leaves. It grows well in drained soils. Tea is a labour-intensive crop. T requires abundant, cheap and skilled labour. It is processed within the tea garden. Major tea producing states are Assam. West Bengal, Tamil Nadu and Kerala. India is the leading producer as well as exporter of tea in the world.

Coffee is the second most important beverage crop of India. It requires hot and humid climatic conditions. Temperature between 15 and 28°c and rainfall between 150 and 200 cm are necessary for growing coffee. Its cultivation is confined to hills around Nilgiris, Karnataka, Kerala and Tamil Nadu are major producers of coffee. Indian coffee is known for its quality and, hence, is in much demand in the international market. It brings in a handsome amount of foreign exchange.

### Cash Crops

*Sugarcane*: It is the main source of sugar, *gur* and Khandsari and is a long-duration crop. Hot and humid climate with a temperature of 21 to 27°C and a rainfall of 75 to 100 cm is idea for the crop. Irrigation is required in areas of low rainfall. About 92 per cent of sugarcane lands are under irrigation. The crop grows in a variety of soils and needs manual labour in large number in various processes of cultivation. India is the second largest producer of sugarcane after Brazil in the world. The major sugarcane producing states are Uttar Pradesh, Maharashtra, Karnataka,Tamil Nadu and Andhra Pradesh.

*Rubber*: Its produced from the latex of a number of different species of rubber tree. It is used for the manufacture of a number of industrial products, including tyres and tubes. The tree requires hot and humid climate with enough rain, about 300 cm. India is the fifth major rubber producing country of the world. In India, it is produced in Kerala, Tamil Nadu, Karnataka, Assam and Tripura. Kerala produces about nine-tenths of the total production. About 97 per cent of the country's demand of natural rubber is met from the domestic production.

*Tobacco*: India is an important producer of tobacco in the world. It is an important cash crop of the country. Besides its use for smoking and chewing. It is also used for manufacturing insecticides. It is a tropical and subtropical plant and, hence, grows in a wide range of temperature from 16 to 35°C. It grows well in areas of 100 cm of rain. In areas of lower rainfall. Irrigation is necessary, Frost is injurious to its growth. For its cultivation. Soil fertility is more important than the climate. It is labour-intensive crop. Major tobacco producing states are Gujarat, Uttar Pradesh, Andhra Pradesh and Karnataka. Abut one-fifth of the production is exported. Major buyers are Russia and the United Kingdom. It is also exported to Japan, Egypt and Sri Lanka.

*Spices*: The spices, by their very name, add flavour and taste to our food. Some of them help in the easy digestion of our food. A larger number of items are covered under this category. They include pepper, cardamom, cinnamon, clove, ginger, onion, garlic, chilli and turmeric. Pepper and cardamom are the products of the equatorial type of climate. As such, Indonesia is the world's largest producer and exporter of these spices, closely followed by India. In India, Kerala leads in production and export of these spices. Followed by Karnataka and Tamil Nadu.

*Fruits*: India is the world's second largest producer of fruits and vegetables. It leads the world in the production of mangoes and bananas, Important fruits grown in India are mango, banana, apple, cashew nut, orange, grapes, peach, pear and strawberry. Important producers of cashew nut are Kerala. Tamil Nadu and Andhra Pradesh. India is the largest exporter of cashew nut in the world.

Mango is the most important fruit, covering about 38 per cent of the area under fruits. It accounts for 23 per cent of the total fruit production. Mango is grown widely in almost all parts of the country. Apple is a temperate fruit crop, hence, it grows mainly in Himachal Pradesh, Kashmir Valley of Jammu and Kashmir and Uttaranchal.

Banana is a tropical as well sub-tropical plant. It grows well in areas of 150 cm or more rain. Though its cultivation is well spread in the country. Tamil Nadu and Maharashtra produce more than the half of the total. Orange is an important citrus fruit. Maharashtra, Uttaranchal. Himachal Pradesh, West Bengal, Meghalaya, Andhra Pradesh, Karnataka, Kerala and Tamil Nadu are important producers of oranges in India.

Grapes, a sub-tropical plant, requires long summer and short winter. In north India, only it is two crops a year. Major producers are Utaranchal, Himachal Pradesh, Jammu and Kashmir, and Punjab in the north, and Mharashtra, Andhra Pradesh, Tamil Nadu and Karnataka in the south.

India produces about 13 per cent of the world's vegetable output, occupying first position in the production of pea and cauliflower, second in onion, cabbage, tomato and brinjal, and fourth rank in potato.

**Animal Husbandry**

For farmers and farms, farm animals form an important ecosystem in an agricultural country like India. Farm animals are the partners of the farmers. Ox, he buffalo and camel are used as draught animals in performing activities like ploughing, sowing, thrashing and transporting farm products. Cows and she buffalos provide them milk.

India has about 57 per cent of the world's buffalow population and 16 per cent of the cattle population. India is now the largest producer of milk in the world which is largely due to White Revolution. Madhya Pradesh, Uttar Pradesh. Bihar, Maharashtra, Orissa, Karnataka and Rajasthan have over two-thirds of the cattle population of India. Andhra Pradesh, Tamil Nadu, Assam and Gujarat also have considerable number of cattle. Crossbreed from foreign and native cattle have also been developed. Buffaloes are reared mainly India. Uttar Pradesh alone has about one fourth of the country's buffaloes. Other important states known for buffaloes are Andhra Pradesh, Madhya Pradesh, Rajasthan, Maharashtra, Bihar, Gujrat, Punjab, Haryana and Karnataka. A number of cattle development projects have been initiated by the government for cattle and buffaloes.

Sheep are raised in those areas that are normally not suitable for agriculture and cattle rearing. One-fourth of the total sheep of India is found in Rajasthan. Andhra Pradesh. Tamil Nadu, Karnataka, Maharashtra, Jammu and Kashmir and Uttar Pradesh share about 50 per cent of the country's sheep. Sheep provide us wool, mutton and skin. Goat is a poor man's cow. It supplies milk, meat, hair, hides and skin. More than half of India's goats are found in Bihar, Jharkhand, Rajasthan, West Bengal Jharkhand, Rajasthan, West Bengal and Uttar Pradesh. The term poultry refers to domestic fowls which are reared for eggs and feathers.

*Fisheries:* Fish forms an important source of protein. Thee are two types of fisheries—marine and freshwater. The former is obtained from coastal areas and the latter from rivers, lakes, ponds and reservoirs. Fisheries sector plays a important role in socio-economic development of the country as it is an important source of livelihood of the people.

# 2

# Towards Agricultural Development in India

India is an agricultural country. Nearly two-thirds of its population depends directly on agriculture for its livelihood. Agriculture is the main stay of India's economy. It accounts for 26 per cent of the gross domestic product. It ensures food security for the country and produces several raw materials for industries. Agricultural development, therefore, is a precondition of our national prosperity. Agriculture is the mainstay of the Indian economy. Agriculture and allied sectors contribute nearly 22 per cent of Gross Domestic Product (GDP of India), while about 65-70 per cent of the population is dependent on agriculture for their livelihood. The agricultural output, however, depends on monsoon as nearly 60 per cent of area sown is dependent on rainfall. During 2004, the southwest monsoon was erratic with prolonged weak/break monsoon conditions prevailing during late June, most part of July, late August and early September over different parts of the country. For the country as a whole, the seasonal rainfall from 1 June to 30 September was 87 per cent of its long period average (LPA). According to the Third Advance Estimates on production of food grains and commercial crops released on 23 March 2005, the production of foodgrains during 2004-05, is estimated at 210.44 million tonnes, which is less than the foodgrains production of 2003-04 by 3.02 million tonnes or 1.4 per cent. The estimated production of rice at 87.12 million tonnes is lower by 1.16 million tonnes or 1.3 per cent as compared to 2003-04. The production of wheat during 2004-05 estimated at 74.05 million tonnes exceeds last year's production by 1.94 million tonnes or 2.8 per cent. The production of coarse cereals estimated at 34.26 million tonnes is less than the last year's production by 3.86 million tonnes or by 10.1 per cent. The production of bajra is estimated at 6.93 million tonnes indicating a decrease of 42.8 per cent over the last year's production of 12.12 million tonnes. In case of maize the production is estimated at 14.87 million tonnes which is less than last years production by 0.06 million tonnes or 0.4 per cent. The expected production of total pulses at 15.01 million tonnes shows 0.5 per cent increase over the last year's production of 14.94 million tonnes.

The estimated production of total oilseeds at 25.15 million tonnes is less than the last year's production by 0.14 million tonnes or by 0.6 per cent. The production of cotton in 2004-05 is estimated at 16.07 million bales of 170 Kg each which is more than the last year's production level by 2.20 million bales indicating an increase of 15.9 per cent. Jute and Mesta production during 2004-05 is estimated at 9.96 million bales of 180 Kg each which is lower than last year's production of 11.23 million bales 11.3 per cent. Sugarcane production is likely to be 234.67 million tonnes, showing a reduction of 2.64 million tonnes or 1.1 per cent over the preceding year's production of 237.31 million tonnes.

Annual growth of crop area, production and yield in respect of cereals, oilseeds and other crops is given below:

**Table 2.1: Annual Compound Growth of Crop Area, Productionand Productivity**

| | *2002-03* | | | *2003-04* | | |
|---|---|---|---|---|---|---|
| *Crop* | *Area* | *Production* | *Yield* | *Area* | *Production* | *Yield* |
| Rice | -8.3 | -23.1 | -16.1 | 3.2 | 22.9 | 19.1 |
| Wheat | -4.4 | -9.6 | -5.5 | 5.5 | 9.7 | 3.9 |
| Coarse Cereals | -8.6 | -21.9 | -14.1 | 14.1 | ·46.9 | 28.7 |
| Total Cereals | -7.4 | -18.7 | -12.3 | 7 | 21.1 | 13.2 |
| Total Pulses | -6.9 | 12.3 | 20.6 | -6.1 | -7.3 | -1.3 |
| Foodgrains | -7.3 | -14.9 | -8.2 | 4.6 | 16.4 | 11:3 |
| Total Oilseeds | -5.1 | -28.5 | -24.7 | 10.3 | 70.5 | 54.8 |
| Cotton (Lint) | -16 | -12.7 | 3.9 | -0.5 | 58.9 | 59.8 |
| Sugarcane | 2.5 | -3.3 | -5.6 | 0.1 | -17.4 | -17.5 |
| Non-Foodgrains' | -10.9 | -11.1 | -0.2 | 4.9 | 15.6 | 10.2 |
| All Principal Crops' | -7.8 | -13.2 | -5.8 | 4.7 | 16.1 | 10.8 |

• Provisional

Based on Index Numbers with Triennium ending 1993-94= 100

## Land Utilisation

Land utilisation statistics are available for 93.1 per cent of total geographical area of 3,287.3 lakh hectares. According to Land Use Statistics available from States, area under forest had increased from 404.8 lakh hectares in 1950-51 to 694.1 lakh hectares in 2000-2001. Net sown area increased from 1,187.5 lakh hectares to 1411.0 lakh hectares during the same period. Broad cropping pattern indicates that though foodgrains have a preponderance in gross cropped area as compared to non-food grains, their relative share came down from 76.7 per cent during 1950-51 to 65 per cent during 2000-2001. According to the Agriculture Census, the area operated by large holdings (10 hectares and above) has declined to 14.8 per cent in 1995-96 compared to 17.3 per cent in 1990-91. Similarly, the area operated under marginal holdings (less than one hectare) has increased to 17.2 per cent in 1995-96 from 15.0 per cent in 1990-91. This indicates that land is getting fragmented.

## Technology Mission on Oilseeds, Pulses and Maize

The Technology Mission on Oilseeds was launched by the Central Government in 1986 to increase the production of oilseeds to reduce import and achieve self-sufficiency in edible oils. Subsequently, pulses, oilpalm and maize were also bi-ought within the purview of the Mission in 1990-91, 1992 and 1995-96 respectively. In addition, the National Oilseeds and Vegetable Oils Development (NOVOD) Board is also supplementing the efforts of TMOP by opening of newer areas for non-traditional oilseeds. It is promoting tree-borne oilseeds. The schemes implemented under TMOP are

(i) Oilseeds Production Programme (OPP);

(ii) National Pulses Development Project (NPDP);

(iii) Accelerated Maize Development Programme (AMDP);

(iv) Post Harvest Technology (PHT);

(v) Oil Palm Development Programme (OPDP); and

(vi) National Oilseeds and Vegetable Oils Development Board (NOVOD).

During the Tenth Plan, these schemes have been restructured as

(i) Integrated Scheme of Oilseeds, Pulses, Oil Palm and Maize (ISOPOM) by converging the ongoing schemes of OPP, NPDP, AMDP & OPDP;

(ii) Research and Development in· Post Harvest and Processing Technology (PH1) in Oilseeds, Pulses, Oil Palm and Maize including TMOP Headquarters; and

(iii) Integrated Development of Tree Borne Oilseeds to be implemented by NOVOD Board.

***Integrated Scheme of Oilseeds, Pulses, Oil Palm and Maize***. In order to provide flexibility to the States in implementation based on regionally differentiated approach, to promote crop diversification and to provide focused approach to the programmes, the four on-going schemes of OPP, OPDP, NPDP and AMDP have peen merged into one Centrally Sponsored Integrated Scheme of Oilseeds , Pulses, Oil Palm and Maize (ISOPOM) during the $10^{th}$ Five year Plan which is being implemented with effect from 1 April 2004.The scheme is implemented by 14 major growing States for oilseeds and pulses and 15 States for Maize and in 1.0 states for oil palm.

The ISOPOM has the following special features

(i) Flexibility to the States to utilise the funds for the scheme/crop of their choice;

(ii) Annual action plan to be formulated by the State Governments for consideration and approval of the Government of India;

(iii) Flexibility to the States for introducing innovative measures or any special component to the extent of 10 per cent of financial allocation;

(iv) Involvement of private sector by the State Governments for the implementation of the programme with a financial cap of 15 per cent.;

(v) Flexibility for inter component diversion of funds upto 20 per cent for non-seed components only; and

(vi) Diversion of funds from seed components to non-seed components with the prior approval of the Department of Agriculture and Cooperation.

New components of ISOPOM included during, $10^{th}$ Plan are:

(i) Inclusion of HOPE pipes;

(ii) Publicity;

(iii) Involvement of Private Sector including NGOs, Farmers' Organisation, Cooperative bodies, Public Sector Agencies, etc;

(iv) Innovative measures and additional components;

(v) Training of Officers/Extension workers; and

(vi) Foreign Visits.

Implementation of oilseeds production programme has helped in increasing the production of oilseeds from 108.30 lakh tonnes in 1985-86 to 252.90 lakh tonnes during 2003-04. The pulses production in the country increased from 128.60 lakh in 1989-90 to 149.40 lakh tonnes in 2003-04. The area under pulses increased from 234.10 lakh ha. in 1989-90 to 244.50 lakh ha. in 2003-04. The area under oil palm increased from 8585 ha. at the end of 1992-93 to 44789 ha. by the end of the $9^{th}$ Plan (2001-2002). Actual production of Fresh Fruit Bunches (FFBs) at the end of the $9^{th}$ Five Year Plan (2001-2002) was 1,28,873 tonnes, yielding around 22,613 tonnes of crude palm oil

(CPO). The production of maize increased from 88.84 lakh tOlmes in 1994-95 to 149.3 lakh tonnes in 2003-04 and the area under maize increased from 61.36 lakh ha. in 1994-95 to 73.22 lakh ha. in 2003-04.

*UNDP Sub Programme on Maize-based Cropping System for Food Security in India.* A Sub-programme on maize based cropping system for food security in India under GOI-UNDP Food Security programme for a total amount of US $ 8,14,000 being fully funded by UNDP, for implementation in six districts in three States viz., Bihar, Rajasthan and Uttar Pradesh for four years, w.e.f. 1999-2000, has also been undertaken.

*Research and Development in Post Harvest and Processing Technology in Oilseeds, Pulses, Oil Palm´ and Maize.* The scheme of Post Harvest Technology (PHT) for Oilseeds, Pulses and Maize was brought under the purview of the TMOP&M with the objective of stepping up their production through efficient scientific processing for realising the ultimate objective of self-reliance in edible oils and other crops and bringing about a position of zero imports.

The Technology Mission has been emphasising on development of Post Harvest Technology (PHT) with the following objectives; (i) Optimising oil recovery from oilseeds by developing efficient modern oil expellers; (ii) Tapping full potential of Rice Bran Oil (RBO) — in the country; (iii) Developing value added products in respect of oilseeds, pulses and maize so that farmers are able to get a better/remunerative price; (iv) Designing and developing indigenous technology for processing of non-conventional and tree borne oilseeds including oil palm; (v) Developing of indigenous technology in maize processing suitable in the context of Indian situations; and (vi) Dissemination of the R&D Technologies developed through demonstration programmes.

The indigenous technologies developed through R&D programmes are popularised by setting up demonstration, units in industries with partial grant-in-aid assistance from TMOP&M, Department of Agriculture and Cooperation. The partial grant-in-aid assistance is provided in the form of plant and machinery.

### Integrated Development of Tree Borne Oilseed by NOVOD Board.

National Oilseed and Vegetable Oils Development (NOVOD) Board is exploring and augmenting the potential of Tree Borne Oilseeds (TBOs) in the country by undertaking activities like back-ended credit linked subsidy programmes consisting of establishment of seed procurement centres, installation of multi-purpose pre-processing and processing facilities, installation of oil expellers, nursery raising and commercial plantation and promotional programmes like development of elite planting material, elite nursery raising and plantation, awareness programmes, establishment of TBO garden, etc

The Board has initiated plantation programmes of *Jatropha curcas,* and Karanja

*(Pongamia pinnata)* for bio-diesel production in the country in more than 10,000 ha. areas in the 25 States/Union Territories of the country. Besides 128 TBO gardens have also been established to demonstrate the silvicultural practices coupled with IPM for early adoption of TBO plantation in 12 States. Besides transfer of technology programmes consisting of 136 farmers' training, 48 trainers' training and 22 workshops have also been organised for motivating farmers to adopt TBO plantation and create awareness among primary seed collectors. The elite planting material and planting techniques as well as location specific package of practices are being evolved through a national network of 42 R&D institutions in the country.

Under the awareness programme, 2300 farmers have been trained and motivated for raising TBOs by conducting 46 farmers' training in the various States_ Similarly, 600 field level functionary and volunteers have been trained through 12 trainers' training programme.

Eight oil expellers have been installed in the State of A.P., Bihar, Karnataka and Tamil Nadu. Six more oil expellers are being installed for extraction of oil from TBOs.

## Crop Production Programme

The Crop Development schemes, viz.,

(i) Integrated Cereal Development Programme in Wheat Based Cropping Systems Areas (ICDP-Wheat),

(ii) Integrated Cereals Development Programme in Rice Based Cropping Systems Areas (ICDP-Rice);

(iii) Integrated Cereal Development Programme on Coarse Cereals Based Cropping Systems Areas (ICDP-Coarse Cereals),

(iv) Sustainable Development of Sugarcane Based Cropping System (SUBACS); and

(v) Special Jute Development Programme (SJDP) implemented during the Ninth plan have since been subsumed under Macro Management Scheme w.e.f. October 2000 to give more flexibility to the State Governments. Assistance is provided to the States for implementation of these programmes based on the Work Plans prepared by them according to local needs. The Central Sector Minikit Programmes of Rice, Wheat and Coarse Cereals have been discontinued from 2002-03.

The Centrally Sponsored Scheme of Intensive Cotton Development Programme (ICDP) which was merged with the Technology Mission on Cotton, would be continued during 2005-06. This scheme is being implemented in 13 States with collaboration between Ministry of Agriculture and Ministry of Textiles. The aim of the scheme is to increase the production and quality of cotton. The scheme provides assistance for extension and development activities.

As a new initiative, a Centrally Sponsored Scheme - On Farm Water Management for encouraging crop production in Eastern India has been launched in March 2002. The objective of the scheme is to increase the production and productivity of foodgrains through focus on exploitation of ground/surface water and its efficient utilisation. The scheme will help in enhancing the income of the farmers and diversifying agriculture production in eastern India. The Scheme is in operation in all districts of Assam, Arunachal Pradesh, Bihar, Chhattisgarh, Jharkhand, Manipur, Mizoram, Orissa, nine districts of West Bengal and 35 districts of Eastern Uttar Pradesh. The main components of the scheme are:

(i) assistance for installation of shallow tubewells with pump sets;

(ii) assistance for electric/diesel pump sets;

(iii) assistance for community lift irrigation points; and

(iv) assistance for dug wells in the plateau region.

The scheme is being implemented on a credit linked back-ended subsidy basis through NABARD in coordination with the State Governments. The funding pattern of the scheme is 20:30:50 basis i.e., 20 per cent contributed by the beneficiaries, 30 per cent subsidy from Government of India and 50 per cent as bank loan.

## Horticulture

India, with its wide variability of climate and soil, has good potential for growing a wide range of horticulture crops such as fruit, vegetables, potato, tropical tuber crops and mushrooms, ornamental crops, medicinal and aromatic plants, spices and plantation crops like coconut, cashewnut, cocoa, etc. Since the mid-eighties, the Government identified horticultural crops as, a means of diversification for making agriculture more profitable through efficient land use, optimum utilisation of natural resources and creating skilled employment for rural masses, especially women folk with the past efforts rewarding. India has emerged as the largest producer of coconut, arecanut, cashew nut, ginger, turmeric, black pepper and the second largest producer of fruits and vegetables. Among the new crops, kiwi, olive crops 'and oil palm have been successfully introduced for commercial cultivation.

The changing scenario encourages private investment to go for hi-tech horticulture with micro-propagation, protected cultivation, drip irrigation, fertigation, and integrated nutrients and pest management, besides making use of latest post-harvest measures, particularly in the case of perishable commodities. As a result, horticulture crop production has moved from rural confines to commercial ventures.

### *Fruits*

A large variety of fruits are grown in India, of those, mango, banana, citrus, pineapple, papaya, guava, sapota, jackfruit, litchi, and grape, among the tropical and sub-tropical fruits; apple, pear, peach, plum, apricot, almond and walnut among the temperate fruits; and aonla, ber, pomegranate, annona, fig, phalsa among the arid zone fruits are important. India accounts for about 10 per cent of the production of fruits in the world. It leads the world in the production of mango, banana, sapota and acid lime and has recorded highest productivity in grape. Mango is the most important fruit covering about 39 per cent of the area and accounts for 23 per cent of total fruit production in the country. India's share in the world production of mango is about 54 per cent. Citrus ranks second in area and accounts for about 10 per cent of total fruits in the country. Limes, lemons, sweet orange and mandarin, cover bulk of the area under this group of fruits. Banana ranks third in area covering about 13 per cent of the total area. India occupies the first position in banana production of 1.16 million tonnes. The percentage share of production of fruits like guava, papaya is about four per cent and litchi is about one per cent. The arid zones of the country are potential areas for fruits like aonla, ber, pomegranate, annona, etc. There has been a steady increase in the area and production of these fruits particularly aonla, ber and pomegranate in the country as a result of identification and development of suitable varieties and production technologies. In addition to these, date palm and fig cultivation are also finding favour in suitable areas.

### *Vegetables*

More than 40 kinds of vegetables belonging to different groups, namely, solanaceous, cucurbitaceous, leguminous, cruciferous (cole crops), root crops and leafy vegetables are grown in India in tropical, sub-tropical and temperate regions. Important vegetable crops grown in the country are tomato, onion, brinjal, cabbage, cauliflower, okra and peas. India is next only to China in area and production of vegetables. India contributes about 13 per cent to the world vegetable production and occupies first position in the production of cauliflower, second in onion and third in cabbage in the world.

### *Spices*

Spices constitute an important group of horticulture crops. India is known as the home of spices and produces a wide variety of spices like black pepper, cardamom (small and large), ginger, garlic, turmeric, chilly and a large variety of tree and seed spices. Among various spices grown in the country, chilly is the most widely grown spice with a share in the total production of 33.7 per cent. The demand for chilly as a spice and its oleoresins as a natural colouring material is growing in the domestic as well as international market. Turmeric has a share of 21.6 per cent in the total production of spices.

### *Plantation Crops*

Other than tea, coffee· and rubber, rest of, the crops which are grown for commercial purpose in the country are covered under plantation crops which include coconut, arecanut, cashew nut, cocoa, etc.

Coconut is an important crop and about 10 million people depend on coconut cultivation, processing and related activities. In India, coconut is grown mainly along the coastal States of the country and also in the North Eastern region. The coir obtained from processing coconut husk is of high commercial value. Beside coir, shell based products have also gained entry into the national international markets. The coir pith made into brick like structure is now used for raising horticultural plants especially in greenhouses.

Cashew nut assumes an important place in the Indian economy. India produces 45 per cent of the global production of cashew. Besides, India is the largest producer, processor, consumer and exporter of cashew in the world.

### *Flowers*

Though flower cultivation has been practised in India since time immemorial, floriculture has blossomed into a viable business only in recent years. Considering the potential this sector has in generating income and employment opportunities, promoting greater involvement of women and enhancement *of* exports, it has been identified as an Extreme Focus Area for exports by the Government of India. India is known for growing traditional flowers such as jasmine, marigold, chrysanthemum, tuberose, crossandra and aster. Commercial cultivation of cut flowers such as rose, orchids, gladiolus, carnation, anthurium, gerbera and lilies has also become popular.

### *Medicinal and Aromatic Plants*

This sub-sector has high potential for health management considering the importance being given to herbal products. Concerted efforts have been made to consume the bio-diversity of the herbal wealth by establishing 226 hectares of herbal gardens along with 16 nurseries to provide quality planting material. Besids, 335 hectares have been developed for production of quality planting material of aromatic plants. More than 6,000 demonstration-cum-seed production plots have been established in the farmers' field.

*Bee-Keeping.* In order to maximise horticulture production, honey bees can be used as an important input for increasing the productivity of horticultural crops through cross pollination. About 85 per cent crop plants are cross-pollinated as they need to receive pollen from other plants of the same species with the help of external agents. Honey bees wax and royal jelly give additional income to the farmers.

*National Horticulture Mission:* The Centrally Sponsored Scheme, namely, National Horticulture Mission has been approved by the Cabinet Committee on Economic Affairs with an outlay of Rs 2300 crore, for implementation during the remaining period of Tenth Five Year Plan and, will address the issues of production, post-harvest management and marketing of horticulture produce.

## Macro Management of Agriculture

Agriculture being a State subject, the primary responsibility for increasing agriculture production, enhancing productivity and exploring the vast untapped potential of the sector rests with the States. The role of the Centre is to complement and supplement their efforts in a catalytic way so that the efforts of the States in agricultural development yield quick results and the benefits percolate down to the poorest farmers. The Macro Management of Agriculture Scheme which was evolved by integrating 27 identified schemes is being implemented in all States/Union' Territories since 2000-01. Under this Scheme, the States have been given flexibility to develop and pursue activities on the basis of their regional priorities.

The new approach ensures that the funds placed at the disposal of the States are not diverted for non-developmental and non-agriculture purposes and are utilised for development of agricultural through specific interventions. The Central assistance is released in lump sum on the basis of the proposals contained in the Work Plans submitted by the various State Governments/Union Territory Administrations. The State Agriculture Department which is the Nodal Department for the Macro Management Scheme further allocates funds to the other Departments/implementing agencies in their States/UTs. The position regarding Budget Estimates, Revised Estimates and expenditure incurred under the scheme since 2000--01 is as indicated below :

*(Rs. in Crore)*

| *Year* | *Budget Estimate* | *Revised Estimate* | *Expenditure* |
|---|---|---|---|
| 2000-01 | 490.00 | 381.88 | 381.88 |
| 2001-02 | 850.00 | 680.49 | 678.62 |
| 2002-03 | 736.86 | 597.00 | 597.59 |
| 2003-04 | 700.00 | 648.60 | 648.49 |
| 2004-05 | 712.92 | 1189.20 | 1188.94 |

## Natural Resources Management

Soil and water conservation measures are one of the essential inputs for increasing agricultural output in the country. These programmes were first launched during the First Plan. From the very beginning, emphasis has been on development of technology for problem identification, enactment of appropriate legislation and constitution of policy coordination bodies. While conceptual framework of soil and water conservation activities has been changed, concept of programmes has undergone considerable revision during successive Five Year Plans.

The Centrally-Sponsored Scheme of Soil Conservation in the Catchments of River Valley Project (RVP) was started in Third Five Year Plan. Subsequently, another scheme of Flood Prone Rivers (FPR) was started in the Sixth Five Year Plan keeping in view the magnitude of floods in the year 1978. Both the schemes were clubbed together during Ninth Five Year Plan on recommendation of Expenditure Finance Committee and were subsumed under Macro Management Mode since November'200D. Under the programme for the Catchment Management of River Valley Projects and Flood Prone Rivers, 53 catchments are covered, spread over 27 States. The total catchment area is 112.03 m. ha. with Priority Area needing urgent treatment in 26 m. ha. Out of this, 5.86 m.ha. have been treated till 2004--05 with an expenditure of Rs 1839.0 crores.

A Centrally-Sponsored Scheme of Reclamation of Alkali Soil was taken up in Punjab, Haryana and Uttar Pradesh during the Seventh Five Year Plan. During the Ninth Plan, extension of the Scheme to all other States of India was approved where alkali soil problems exist as per scientific parameters. The Scheme aims at improving physical conditions and productivity status of alkali soils for restoring optimum crop production. The major components of the Scheme include, assured irrigation water on farm development works like land leveling, bunding and ploughing, community drainage system, application of soil amendment organic manures, etc. An area of 0.64 m. ha., out of 3.5 m. ha., of alkali land has been reclaimed **till** the end of 2004-05 in the country. The Scheme at present stands subsumed within the Macro Management Scheme.

The Scheme of Watershed Development Project in Shifting Cultivation Areas (WDPSA) was launched in seven north-eastern States during the Eighth Plan from 1994-95 with 100 per cent Central assistance to the State Plan. The Scheme aims at overall development of jhum areas on watershed basis. During the Eighth Plan an amount of Rs 40.826 crore was released to the States of north-east region and 0.67 lakh ha. area was treated through treatment packages. During the Ninth Plan, 1.57 lakh ha. have been treated with an expenditure of Rs 82 crore (against the approved programme and unspent balance of Eighth Plan). The new guidelines of the scheme on the basis of new watershed to common approach has been effective from November 2000 in the revised cost norms of Rs 10,000 per ha. on net treatable area basis with additional activities and improved institutional mechanism. During the Tenth Plan, an area of 0.6 lakh ha has been treated at an expenditure of Rs 60 crore up to end of 2004-05.

## Rainfed Farming

Rainfed Farming is complex, diverse and risk prone and is characterised by low levels of productivity and low input uses. The Government of India has accorded high priority to the holistic and sustainable development of rainfed areas through integrated watershed development approach. The key attributes of the watershed approach are conservation of rain water and optimisation of soil and water resources in a sustainable and cost--effective mode. Improved moisture management increases the productivity of improved seeds and Fertilizer. So conservation and productivity enhancing measures become complementary

A National Watershed Development Project for Rainfed Area (NWDPRA) was launched in 1990-91 in 25 states and two UTs. The project was continued during Ninth Five Year Plan in 28 states inclusive of three newly created States and two UTs with the purpose of increasing agriculture productivity and production in rainfed areas. The guidelines for NWDPRA have been radically restructured by retaining the technical strength and incorporating community participation. The watershed development programme is now planned, implemented, monitored and maintained by watershed communities. During the Ninth plan, an area of 27.66 lakh ha. was treated. Now the Scheme of NWDPRA has been subsumed within the Macro Management of Agriculture Scheme. The Scheme is being continued for implementation during the Tenth Plan under the revised guidelines, with people's participation, covering more than 6000 micro-watersheds. It is estimated to develop an area of about 20 lakh ha during Tenth Plan.

A Watershed Development Fund (WDF) has been established at NABARD with the objective of integrated watershed development in 100 priority districts of 18 States through participatory approach. The total corpus of WDF is Rs 200 crore. Under WDF, two-thirds of amount is given for loan based project and one-third of amount is given for grant based project in the State. A number of externally aided projects are also under implementation on watershed approach, which covers an area of about 1.5 lakh hectares anm;ally.

## Integrated Nutrients Management

The main concern of Integrated Nutrients Management( INM )Division is to ensure adequate availability of quality fertilizers to farmers through periodical demand assessment and timely supply, promoting soil test based judicious and balanced use of chemical fertilizers in conjunction with organic manures and bio-fertilizers, promotion of organic farming and ensuring quality control of fertilizers through implementation of Fertilizer (Control) Order, 1985, to give additional thrust on integrated use of all sources of plant nutrients through chemical fertilizers and organic manure, etc.

India is the third largest producer and consumer of fertilizers in the world after China and USA. Against 16.8 million tonnes of fertilizer nutrients (NPK) consumed during 2003-04, the nutrient consumption is estimated at 18.7 million tonnes during 2004-05. The consumption of major fertilizers namely, Urea, DAP, MOP, SSP and Complexes are estimated to be 20.9, 6.5, 2.4, 2.7 and 5.68 million tonnes respectively during 2004-05. India is 100 per cent self sufficient in respect of urea and about 95 per cent in case of OAr. All-India average fertilizer consumption is 89.9 kg./ ha though there is wide variation from State to State from 184 kg/ha in Punjab, 167 kg/ha in Haryana to less than 10 kg/ha in States like Arunachal Pradesh, Nagaland, Sikkim, etc. Considering the skewed pattern of fertilizer use, Government of India is promoting the balanced and integrated use of Fertilizer nutrients through various initiatives. As a result, NPK consumption ratio has improved to 6.5:2.5:1 during 2003-04 from 7.0:2.7:1 during 2000-01.

*Price of fertilizers and Concession/Subsidy.* The price of urea fertilizer is fixed by the Government. To ensure adequate availability of fertilizers to farmers at reasonable rates, subsidy is provided by Government of India. Urea, the most consumed fertilizer, is subsidised under the New Urea Pricing Scheme, whereas P&K fertilizers, which are decontrolled, are covered under the Concession Scheme. The existing scheme for special freight subsidy has been continued for supplies to the North Eastern States and Jammu & Kashmir. The prices of major fertilizers are as under:

The Government is implementing a Centrally Sponsored Scheme 'Balanced and Integrated Use of Fertilizers' to popularise soil test based judicious application of fertilizers in combination with organic manures and biofertilizers. In addition, financial assistance was provided for setting up/ strengthening of new soil testing laboratories for advising farmers on major and micro nutrients application, training and demonstration on balanced use of fertilizers and also production of useful organic manure from city waste/ garbage.

This scheme has since been subsumed within Macro Management Scheme with effect from October 2000. The State Governments are required to undertake all these activities through their annual Work Plans.

## Central Fertilizer Quality Control and Training Institute

The Government ensures the quality of fertilizers through Fertilizer Control Order (FCO) issued under Essential Commodities Act to regulate the price, trade, policy and distribution of fertilizers in the country. The State Governments are the executive agency to implement the various provisions of FCO. The Order strictly prohibits the manufacture, import and sale of any fertilizer, which does not meet the prescribed standards. The Central Fertilizer Quality Control and Training Institute at' Faridabad and its 3 Regional centres located at Navi Mumbai, Chennai and Kalyani have been set up for inspection and analysis of imported and indigenous fertilizers, giving technical advice and providing training on quality control to State enforcement agencies and analysts. The Institute has also developed a quick testing kit for on-the-spot detection of adulteration in fertilizers.

The FCO has been recently amended to make it more userfriendly and ensuring effective enforcement. The important amendments include provision for repackaging of cut or torn fertilizer bags and reprocessing of damaged fertilizer due to natural calamities and provisions for commercial

trials of provisional fertilizers. Accordingly, National Fertilizers Limited (NFL), Indo Gulf Fertilizers and Shri Ram Fertilizers have been permitted to manufacture neem-coated urea under C1.20 A for commercial trials and *Mis* IFFCO has been permitted to manufacture fortified complex fertilizers with Boron and Zinc. Printing of Maximum Retail Price(MRP) on all fertilizers including urea is now mandatory along with the printing of month and year of manufacture of fertilizer imported.

## National Project on Development and use of Biofertilizers

With a view to promote the use of bio-fertilizers as an environment-friendly and cheaper source of plant nutrients, Government of India had taken up a 'National Project on Development and Use of Biofertilizers' during the Sixth Plan and was continued with the aim of promoting Biofertilizers upto the Ninth Plan. The Scheme provided for setting up National Biofertilizer Development Centre at Ghaziabad and its six Regional Biofertilizer Development Centres (RBDCs) at Jabalpur, Nagpur, Bangalore, Bhubaneswar, Hissar and Imphal for promotion, extension, training and demonstration of biofertilizers. Under the Scheme, assistance for setting up of biofertilizer production units was also provided. Presently, there are around 125 biofertilizer production units in the country with an annual capacity of 18,000 tonnes of different types of biofertilizers and the annual production is about 10,000 tonnes. This scheme has been subsumed under a new Central Sector Scheme "National Project on Organic Farming" from October 2004. To undertake the new activities, the names of NBDC and RBDCs have been changed to National Centre of Organic Farming (NCOF) and Regional Centres of Organic Farming (RCOF).

## National Project on Organic Farming

A new scheme, 'National Project on Organic Farming' has been taken up in October 2004 with an outlay of Rs 57.05 crore for production, promotion, market development of organic farming in the country during the remaining period of the Tenth Plan. The main components of the Scheme are:

(i) Putting in place a system of certification of organic produce.

(ii) Capacity building through service providers.

(iii) Financial support for commercial production units for production of organic inputs like: (a) fruits and vegetable waste compost units; (b) biofertilizers production; and (c) hatcheries for vermi-culture.

(iv) Promotion and extension of Organic farming.

## Plant Protection

Plant Protection continues to playa significant role in achieving targets of crop production. The major thrust areas of plant protection are promotion of Integrated Pest Management, ensuring availability of safe and quality pesticides for sustaining crop production from the ravages of pests and diseases, streamlining the quarantine measures for accelerating the introduction of new high yielding crop varieties, besides eliminating the chances of entry of exotic pests and for human resource development including empowerment of women in plant protection skills.

Presently, two Central Sector Plan Schemes are being implemented to achieve the main objectives of Plant Protection strategy in the country.

(i) *Promotion of Integrated Pest Management.* Keeping in view the ill effects of chemical pesticides such as development of pest resistance, pest resurgence, outbreak of secondary pests, pesticide residues in food, fodder, soil, air and water resulting in human health hazards and ecological imbalances, Government of India has adopted Integrated Pest Management (IPM) as the cardinal principle and main plank of plant protection strategy in the country. IPM is an eco-friendly approach aimed at minimum use of chemical pesticides by employing available alternative methods for pest control like cultural, mechanical and biological and use of bio-pesticides.

To enhance the production of bio-control agents/bio-pesticides, Government of India earmarked Rs 1372 lakh as grants-in-aid during the Eighth and Ninth Plans for establishment of 29 State Bio-control Laboratories.

With the adoption of IPM approach, consumption of pesticides has come down from 72,133 MT (tech. grade) during 1991-92 to 41,020 MT (tech. grade) during 2003-04 and there is consistent increase in use of bio-pesticides which are environmentally safe. In some of the States, farmers have stopped using chemical pesticides and have adopted organic farming.

(ii) *Locust Control and Research.* The Locust Warning Organisation, Jodhpur with its five locust circle offices and 23 locust outposts monitors two lakh sq. km. Scheduled Desert Area in Rajasthan, Gujarat and Haryana for locust surveillance and control. During 2004-05, an area of 215 lakh hectares has been surveyed and 24 locust bulletins have been issued. Close liaison is being maintained with other locust prone countries and the FAO to keep watch over possible locust invasion. The 24$^{th}$ Session of FAO Commission for Controlling the Desert Locust in South West Asia was held in New Delhi from 10-14 January 2005.

*(iii) Training in Plant Protection.* Until 1966, there was no national training facility in India to cater to the needs of the States. To bridge this gap, the National Plant Protection Training Institute was established in 1966 at Hyderabad (Andhra Pradesh). The Institute has been recognised as a Regional Training Centre for Plant Protection by Food and Agricultural Organisation of the United Nations and also as an Advanced Centre for training in Plant Protection Technology by the World Bank. The Institute has conducted 878 courses and trained 16,223 officers/officials including 216 foreign trainers upto 2004-05.

*(iv) Implementation of Insecticides Act.* Pesticides are commonly used in crop protection measures for sustaining food production. These are also used for the control of vector borne diseases. As they are meant for killing insect pests and, therefore, toxic by their nature. Their misuse or abuse may result in hazards to human beings, animals and the environment. Therefore, pesticides are required to be used safely and judiciously. Keeping this in view, their import, manufacture, sale and use, etc., are being regulated under the Insecticides Act, 1968 and the Rules' framed thereunder. The provisions under the Act ensure the availability of safe, effective and quality pesticides to the farmers.

The Central Government has constituted the Central Insecticides Board under Section 4 of the Insecticides Act, 1968 to advise the Central and State Governments on technical matters arising out of the administration of this Act. Director General, Health Services is the Chairman with 29 members. The Central Government has constituted a Registration Committee to register pesticides after examining their formulae verifying their efficacy and safety to human beings, animals and environment.

The Government of India has also set up a Central Insecticides Laboratory with the major objectives of pre and post registration verification of the properties, performance and hazards of pesticides and the proposed use claimed by the manufacturers. To supplement the resources of the States/UTs in the analysis of pesticides, two Regional Pesticides Testing Laboratories have also been set up at Chandigarh and Kanpur with an analysis capacity of 900 samples per annum each.

## Plant Quarantine Facilities

Plant Quarantine activities are intended to prevent and control/contain exotic pests and diseases into the country by adoption of suitable domestic plant quarantine measures. In order to achieve this goal, the provisions of the Destructive Insects and Pests Act, 1914 and the Plant Quarantine (Regulation of Import into India) Order, 2003 are being implemented. Besides, the work pertaining to Post Entry Quarantine inspection/surveillance in respect of the identified plants/planting material in pursuance of Plant Quarantine Order, 2003 has been assigned to the Inspection Authorities who are the scientists of the State Agricultural Universities. The export certification and issuance of phytosanitary certificate is undertaken by the Plant

Quarantine Stations. Five Regional Plant Quarantine Stations at New Delhi, Mumbai, Chennai, Kolkata and Amritsar have been strengthened under UNDP Project in view of increased import and export of agricultural commodities under WTO provisions. Out of 33 Plant Quarantine Stations, 10 stations are located at Airports, 10 stations at Seaports and 13 stations at Land Frontiers.

## Monitoring of Pesticide Residues

The subject "Monitoring of Pesticide Residues" has been allocated to the Department of Agriculture and Cooperation by the Cabinet Secretariat. Monitoring of pesticide residues at national level is essential to ensure food safety in the country. Further, it is essential to ensure that our export consignments are not rejected due to presence of pesticide residues. The Scheme has been approved with an outlay of Rs 24 crore during the Tenth Plan.

## Agricultural Mechanisation

Strategies and programmes have been directed towards replacement of traditional and inefficient implements by improved ones, enabling the farmers to own tractors, power tillers, harvesters and other machines, availability of custom services, support services of human resource development, testing and evaluation and research and development. A large industrial base for manufacturing of the agricultural machines has also been developed. Introduction of technologically advanced equipments through extension and demonstration besides institutional credit has also been taken up. Equipments for resource conservation have also been adopted by the farmers.

Under various Government sponsored schemes like Macro Management of Agriculture, On Farm Water Management, Technology Mission for Oilseeds and Pulses, Technology Mission on Horticulture and the Technology Mission on Cotton financial assistance is provided to the farmers for the purchase of identified agricultural implements and machines.

*Farm Machinery Training and Testing Institutes.* Farm Machinery Training and Testing Institutes (FMT&TIs) have been established at Budni (Madhya Pradesh), Hissar (Haryana), Garladinne (Andhra Pradesh) and at Biswanath Chariali (Assam) having capacity to train 5000 personnel annually on various aspects of agricultural mechanisation. These institutes also undertake testing and performance evaluation of agricultural machines including tractors in accordance with the national and international standards. Since inception about 83,483 personnel have been trained and about 2042 machines tested by these institutes till 31 March 2005. During 2004-05 these institutes have trained 5601 personnel and tested 111 machines. The training programmes have been revised for giving more emphasis on crop/area/technology specific issues for optimised training duration for their implementation through Farm Machinery Training and Testing Institutes from 1 April 2005.

*Outsourcing of Training*. This is a new component approved for the Tenth Plan under the Central Sector Scheme "Promotion and Strengthening of Agricultural Mechanisation through Training, Testing and Demonstration" in order to train large number of farmers at nearby places. The training programme shall be arranged through the identified institutions by each State namely State Agricultural Universities (SAUs), Agricultural Engineering Colleges/Polytechnics, etc. This Department has released Rs 73.00 lakh during the year 2004-05 to various State Governments for conducting training programmes at the identified institutions. For the year 2005-06, a target to train 3000 farmers has been kept.

*State Agro Industries Corporations*. Seventeen State Agro Industries Corporations, the Joint Sector Companies have been promoted by the Government of India and by the concerned State Governments. The objectives of these Corporations envisage manufacture and distribution of agricultural machines, distribution of agri-inputs, promotion and execution of agro based industries and providing technical services and guidance to the farmers and others. The Government of India's share in six State Agro Industries Corporations namely: Tamil Nadu, Karnataka, Rajasthan, Gujarat, Uttar Pradesh and West Bengal have been disinvested in favour of the respective State Governments.

*Legislative Framework*. The Dangerous Machines (Regulation) Act, 1983, came into force with effect from 14 September 1983. The Act provides for the regulation of trade and commerce and production, supply and use of products of any industry producing dangerous machines with a view to securing the welfare of persons operating any machine and for payment of compensation for death or bodily injury suffered while operating any such machine. Power threshers used for threshing of the agricultural crops have been brought under the ambit of this Act. Government of India have notified the Dangerous Machines (Regulation) Rule, 1985 laying down the specifications for the feeding chutes and for installation of the power threshers. Apart from power threshers it is proposed to bring chaff-cutter and sugarcane crusher under the ambit of this Act.

*Demonstration of newly developed agricultural/horticultural equipments*. With an objective of introducing new technology in agricultural production system, demonstration of newly developed/improved agricultural and horticultural equipment has been approved from the year 1999-2000 which now forms a component of the restructured Central Sector

Scheme "Promotion and Strengthening of Agricultural Mechanisation through Training, Testing and Demonstration". Under this component, 100 per cent Central grant-in-aid is provided to implementing agencies for procurement and demonstration of new/improved equipments. The Scheme is implemented through State/Central

Government organisations. This component has helped in adoption of new agricultural/horticultural equipments by the farmers. During 2004-05, 2927 demonstrations of different new equipments were conducted benefiting 1,62,944 farmers. During 2005-06, 1500 such demonstrations will be conducted through the implementing agencies.

**Extension Activities**

Agricultural Extension aims at improving technology dissemination of farm technologies to the farmers. The Extension System has undergone several changes over the period, in terms of content and delivery approaches. Agricultural Extension is a State subject, however, the Ministry of Agriculture supplements States' efforts in improving extension services.

The Department of Agriculture and Cooperation, in the Ministry of Agriculture lays down major policy guidelines on extension matters and the Directorate of Extension implements specific programmes and activities in close collaboration with the State Departments of Agriculture, the State Agricultural Universities and other concerned agencies.

Current efforts in improving extension services include the following:

- Providing innovative and decentralised institutional arrangements to make extension system farmer driven and farmer accountable.
- Augmentation of Mass Media Support to Agriculture Extension by providing for location specific broadcasts through 96FM stations of All India Radio, 180 Low Powered Transmitters/High Powered Transmitters, 18 Regional Kendras of Doordarshan and also on DO National Channel.
- Operation of Kisan Call Centres through toll free lines throughout the country to provide expert advice to farmers.
- Providing fee based advisory and other support services to the farmers by training unemployed agriculture graduates in agri business development and establishing agri-clincs.
- Encouraging Public Private Partnership (PPP) in various modes/forms can provide synergistic approach in the extension efforts; and other extension programmes.
- Mainstreaming gender concerns in agricultural policies and programmes.

**Seeds**

Seed is a critical and basic input for enhancing agricultural production and productivity in different agro-climatic regions. Indian seed programme largely adheres to the limited generation system for seed multiplication. The system recognises three generations, namely, breeder, 'foundation and certified seeds and provides adequate safeguards for quality assurance in the seed multiplication chain to maintain the purity of variety as it flows from the breeders to the farmers. The level of certified/ quality seeds distributed to the farmers since 2000-01 is given below:-

*(in lakh quintals)*

| *Year* | *Certified/quality seed distribution* |
|---|---|
| 2001-01 | 86.27 |
| 2001-02 | 91.80 |
| 2002-03 | 93.03 |
| 2003-04 | 109.26 |
| 2004-05 | 127.0 (anticipated) |
| 2005-06 | 138.50 (targeted) |

**Structureof Seed Industry**

Indian seed programme includes the participation of Central and State Governments, Indian Council of Agricultural Research (ICAR), State Agricultural Universities (SAU) system, public sector, co-operative sector and private sector institutions, Seed sector in India consists of two national level corporations, Le., National Seeds Corporation (NSC) and State Farms Corporation of India (SFCI), 13 State Seed Corporations (SSCs) and about 100 major seed companies. For quality control and certification, there are 22 State Seed Certification Agencies (SSCAs) and 101 State Seed Testing Laboratories (SSTLs). The private sector has started to playa significant role in the production and distribution of seeds. However, the organised seed sector particularly for food crops cereals continue to be dominated by the public sector.

*Legislative Framework and Policy.* The Seeds Act, 1966 provides for the legislative framework for regulation of quality of seeds sold in the country. The Central Seed Committee (CSC) and the Central Seed Certification Board (CSCB) are apex agencies set up under the Act to deal with all matters relating ·to administration of the Act and quality control of seeds. The proposed Seeds Bill, 2004 in replacement of the Seeds Act, 1966 was introduced in the Parliament in December 2004 and is currently under consideration of the Parliamentary Standing Committee on Agriculture.

In order to encourage export of seeds in the interest of farmers, the procedure for export of seeds has been simplified. Seeds of various crops have been placed under Open General Licence (OGL) except the seeds of wild varieties, germplasms, breeder seeds, jute seeds and onion seeds which are on restricted list under the new Export and Import Policy 2002-07.

*Schemes of the Seed Division.* (i) This Department have launched a Central Sector Scheme namely, "Development and Strengthening of Infrastructure Facilities for Production and Distribution of Quality Seeds" with an outlay of Rs 159 crore for the Tenth Plan. The main components of the scheme are Quality Control Arrangements on Seeds, Transport subsidy on movement of seeds to North-East and other hilly areas, Establishment and Maintenance of Seed Bank, Seed Village Scheme, Assistance for creation of infrastructure facilities, Assistance for boosting seed production in private sector, Human Resources Development, Assistance for Seed Export, Propagation of application of biotechnology in agriculture, Promotion of use of hybrid seeds of rice and evaluation/review. (ii) Legislation on Plant Varieties Protection and Farmers' Rights: In order to fulfil the obligations under TRIPS Agreement of the World Trade Organisation (WTO), which India has ratified, the Department of Agriculture and Cooperation have enacted a legislation for Protection of Plant Varieties and Farmers' Rights. In order to provide necessary back-up support for enactment of the above Legislation, a Central Scheme for implementation of PVP Legislation has been formulated. The required rules and regulations under the Plant Varieties and Farmers' Rights Act have been notified in 2003. The Protection of Plant Varieties and Farmers' Rights Authority envisaged under the Act will be set up shortly.

## Agricultural Marketing

*The Directorate of Marketing and Inspection (DMI)* extends support to the Central and State Governments on agricultural marketing policies and programmes and implements various schemes under them. It has its Head Office at Faridabad, Branch Head-Office at Nagpur, 11 Regional offices, and the Central Agmark Laboratory at Nagpur, 26 Sub-offices and 17 Regional Agmark laboratories spread all over the country.

The DMI administers the Agricultural Produce Grading and Marking Act, 1937 as amended in 1986. The grade standards notified under the Act are popularly known as Agmark standards and such standards have been formulated and notified under the Act in respect of 181 agricultural and allied produce. As a result of special efforts initiated to increase grading under Agmark, 505 new packers were enrolled during 2004-2005. During

2004-2005, agricultural commodities including fruits and vegetables valued at Rs.103.00 crore were graded for export. During this period, agricultural commodities worth Rs 4500.00 crore were graded under Agmark for domestic trade.

The DMI undertakes marketing, research, surveys and in-depth studies of marketing system for various agricultural commodities in the country and suggests remedial measures. The Directorate has conducted country level studies on estimation of marketable surplus and post-harvest losses of 12 foodgrains. Reports on all 12 foodgrains viz., Paddy, Wheat, Jowar, Bajra, Maize, Red gram, Lentil, Green gram, Ragi, Bengal Gram, Barley and Black gram have been prepared and are in the process of printing in final stages. So far, DMI has brought out 330 reports on marketing of various agricultural and allied commodities. Under Research Grants Scheme, 33 rescan studies have been carried out on subjects related to agricultural marketing.

The Directorate is implementing a Central Sector Scheme 'Marketing Research and Information Network' to establish a nationwide network for speedy collection and dissemination of market information. Under this scheme, 993 markets have already been provided connectivity out of which, 764 markets were made operational by the end of March 2005. Besides, connectivity to 411 additional agricultural markets are under process. At present, daily prices of more than 300 commodities and about 2000 varieties are being reported.

With a view to create scientific storage capacity with allied facilities in rural areas to meet the requirements of farmers for storing farm produce, processed farm produce, consumer articles and agricultural inputs, promotion of grading, standardisation and quality control of agricultural produce to improve their marketability, prevention of distress sale immediately after harvest by providing the facility of pledge financing and marketing credit and. to strengthen agricultural marketing infrastructure in the country, another Central Sector Scheme/Construction of Rural Godowns (Gramin Bhandaran Yojna)' is also being implemented by the Directorate. Under the Scheme, 15 per cent/ 25 per cent/33.33 per cent back-ended subsidy (as the case may be) on the project costs is being provided for the purpose. The Scheme which was initially introduced for two years, was extended upto 30 September 2004 and the same has been extended further w.e.f. 1 October 2004 for the remi'lining period of the Tenth Plan i.e., upto 31 March 2007 with certain modifications. Under the Scheme, so far 8061 projects involving storage capacity of 122.63 lakh tonnes have been sanctioned by NABARD/ NCDC for creation of rural godowns. Besides, 1422 projects with a capacity of 19.20 lakh tonnes have been sanctioned by NCDC for renovation of cooperative godowns. Thus, for construction and renovation, 9483 godowns with a capacity of 141.83 lakh tonnes have been sanctioned.

A new Central Scheme viz., "Development/Strengthening of Agricultural Marketing infrastructure, Grading and Standardisation" has recently been sanctioned (on 20 October 2004) and the same is under implementation. It aims at providing credit-linked back-ended subsidy on capital investment for setting up of marketing infrastructure & support services such as grading and standardisation, etc. The Scheme will provide for setting up of new general or commodity specific agriculture produce markets and for strengthening of existing markets, wholesale, rural periodic or in tribal areas to the farm producers any or all of the different categories of services like: (i) assembling, grading, packing, standardisation, quality control and certification; (ii) retailing, wholesaling and e-trading ; iii) direct marketing from producers to consumers; (iv) direct integration of processing units with rural producers; and (v) market intelligence, extension and market oriented production planning.

The Scheme covers agriculture and allied sectors including dairy, meat, fisheries and minor forest produce. This Scheme is however, reform--linked and is implemented in those States/Union Territories which amend their APMC Act to permit setting up of agricultural markets in private and cooperative sector, allow direct marketing and contract farming. The Scheme is implemented by DMI. The States of Andhra Pradesh, Himachal Pradesh, Kerala, Tamil Nadu, Madhya Pradesh, Manipur, Nagaland, Punjab, Sikkim and the UT of Andaman and Nicobar Islands have brought about requisite reforms in their respective APMC Acts or have no legal bar to direct marketing, contract farming and setting up of markets in private and cooperative sector and hence, this scheme has already been stated to be implemented in these States. The other States are in process of amending their APMC Acts.

*Ch. Charan Singh National Institute of Agricultural Marketing (NIAM)* started functioning at Jaipur (Rajasthan) with effect from 8 August 1988. The main functions performed by the Institute relate to Training, Research, Consultancy Projects and Education as well as Policy related assignments from th~ Ministry of Agriculture. The Institute has prepared Agricultural Marketing Plans for the development of Agricultural Marketing in a number of States. It is preparing project reports for setting up Terminal Markets. NIAM conducts Post Graduate Programmes in Agribusiness Management and Agri-warehousing and Cool Chain Management which enjoy a very good reputation. The Institute has been imparting training. to senior and middle level executives of Agricultural and Horticultural departments, agro industries, corporations, state marketing boards, Agricultural Produce Marketing Committees and apex level cooperatives,

Commodity Boards, export houses recognised by Agricultural and Processed Food Products Export Development Authority (APEDA), Commercial banks and non-governmental organisations. Besides these, the Institute also imparts training to farmers on marketing management. The Institute is managed by a Governing Body under the Chairmanship of Minister for Agriculture and an Executive Committee under the Chairmanship of Secretary, Department of Agriculture and Cooperation.

*Small Farmers Agri-Business Consortium (SFAC)* was established in 1994 as a sequel to announcement of Finance Minister in his Budget speech of 1992, with a view to support innovative ideas for generating income and employment in rural areas. SFAC was registered as a Society on 18 January 1994 with the Government of India (Department of Agriculture and Cooperation), Reserve Bank of India, NABARD, IDBI, State Bank of India, Oriental Bank of Commerce, and Exim Bank as promoter members; Canara Bank, United Phosphorous Ltd., and NAFED as primary members. In coordination with the State Governments, SFAC has organised the setting up of State level SFACs in 17 States so far. Each of the State SFACs has been provided corpus assistance by SFAC.

SFAC also functions as a Service provider to the Government of India under the (i) Centrally Sponsored Scheme of Technology Mission for Integrated Development of Horticulture in NE States including Sikkim, since extended to Jammu & Kashmir, Himachal Pradesh and Uttaranchal; (ii) Scheme for Setting up of a Network of Agri-Clinics and Agri-Business Centres.

The Board of Management of SFAC is headed by the Agriculture Minister as ex-Officio President and the Secretary to the Government of India, Department of Agriculture and Cooperation as ex-Officio Vice President.

## Cooperation Cooperative Reforms

The Cooperative movement in India traces its origin to the agriculture and allied sector and was evolved as a mechanism for pooling meagre resources of the people with a view to providing them the advantages of economies of scale. After Independence, the cooperatives were considered as part of the strategy for planned economic development. The cooperatives are today at the crossroads of existence particularly in view of the fast emerging scenario of economic liberalisation and globalisation. These institutions in general suffer from resource constraint, poor governance and management, inefficiency and unviability. The cooperative reforms are, therefore, vital to the future of the cooperatives.

*National Policy on Cooperatives.* The Union Government has formulated a National Policy on Cooperatives in consultation with the States. The objective of the national policy is to facilitate all-rowld development of cooperatives in the country and to work as guiding principles for action of the State towards cooperatives. The policy envisages that cooperatives be provided necessary support, encouragement and assistance and to ensure that they work as autonomous, self-reliant and democratically managed institutions accountable to their members.

*Multi-State Cooperative Societies (MSCS) Act, 2002.* The Central Government has taken a lead in providing required autonomy to the cooperatives by enacting MSCS Act, 2002,'replacing the MSCS Act, 1984. It aims at providing functional autonomy and democratic management to the Multi-State Cooperative Societies. Though this Act is applicable to national level cooperative societies/federations and other Multi-State Cooperative Societies, it is expected that it will work as a model Act for reforms in State Cooperative Laws.

*NCDC(Amendment) Act, 2002.* NCDC Act, 1962 has been amended vide NCDC (Amendment) Act, 2002 expanding the mandate of NCDC to include foodstuff, industrial goods, livestock and services in the programmes and activities in addition to the existing programmes. The definition of agriculture produce has been amended to cover wider range of activities by including edible and non-edible oilseeds, cattle feed) produce of horticulture and animal husbandry, forestry, poultry farming, pisiculture and other allied activities to agriculture. Vide the amended Act, definition of industrial goods and livestock have been elaborated covering the products of allied industries in the rural areas and includes any handicraft or rural craft. The livestock includes all animals to be raised for milk, meat, fleece, skin, wool and other by-products. NCDC will be able to provide loan directly to the co-operatives without State/Central Government guarantee on furnishing of security to the satisfaction of the NCDC. So far activities concerning water conservation, animal care/health, disease prevention, agricultural insurance and agricultural credit, rural sanitation/ drainage/ sewerage have been made notified services.

*Amendment to the Constitution in respect of Cooperative.* In spite of the felt need for amendment in the State Cooperative Societies Acts, the pace of reforms in Cooperative legislations by the States is not encouraging. Therefore, in order to ensure the democratic, autonomous and professional functioning of cooperatives, it has been decided to initiate a proposal for amendment to the Constitution. This decision has been taken after careful deliberations with the States in a conference of State Cooperative Ministers held on 7 December 2004. The proposed amendement to the Constitution is purported to address the key issues for empowerment of co-operatives

through their voluntary formation, autonomous functioning, democratic control and professional management.

*Constitution of High Powered Committee.* To review the achievements of Cooperative movement during the last 100 years of cooperative movement and challenges before' it , and to ,suggest ways and means to face them and to give a new direction to the movement, High Powered Committee has been constituted with the following Terms of Reference:

(1) To review the achievements of the co-operatives during the last 100 years.

(2) To identify the challenges being faced by the co-operative sector and to suggest measures to address them to enable the movement to keep pace with the changing socio-economic environment.

(3) To suggest an appropriate policy and legislative framework and changes required in the co-operative legislation in the country with a view to ensure the democratic, autonomous and professional functioning of co-operatives.

*Revamping of Cooperative Credit Structure.* In order to improve the health of rural cooperative, credit institutions, Government of India constituted a Task Force under the Chairmanship of Prof. A. Vaidyanathan for suggesting measures required for improving the efficiency and viability of short-term cooperative credit structure and to suggest an appropriate regulatory framework. The Task Force has submitted its report and the Government has accepted the recommendations of the Task Force initiated steps to consult various stakeholders for finalising the modalities for its implementation. The same Task Force has been as- signed the task to look into the problems of long-term cooperative credit structure and to suggest measures for its revitalisation

## Animal Husbandry

Animal Husbandry and Dairy Development plays a prominent role in the rural economy in supplementing the income of rural households, particularly, the landless and small and marginal farmers. It also provides subsidiary occupation in semi-urban areas and more so for people living in hilly, tribal and drought prone areas where crop output may not sustain the family. Animal husbandry output constitutes about 30 per cent of the country's agricultural output.

India is endowed with the largest livestock population in the world.

It accounts for 57 per cent of the world's buffalo population and 15 per cent of the cattle population. According to Livestock Census (2003), the country has about 18.5 crore cattle and 9.8 crore buffaloes.

## Contribution of Livestock Sector to Food Basket

The contribution of livestock sector to the food basket in the form of milk, eggs and meat has been immense in fulfilling the animal protein requirement of ever-growing human population. The present availability of human protein in an Indian diet is 10 gm per person per day, as against a world average of 25 gm. However, keeping in view the growing population, the animal protein availability has to increase at least two-fold for maintaining the nutritional level of growing children and nursing mothers in India.

*Milk Production.* During past five year plans, several measures have been initiated by the Government to increase the productivity of livestock, which has resulted in significant increase in the milk production to the level of 88.1 million tonnes at the end of 2003-04. India has become the largest producer of milk in the world. The per capita availability of milk is estimated to be 231 gm per day during 2003-04.

*Egg Production.* Poultry development in the country has shown steady progress over the years. Egg production during 2003:04 was 40.4 billion. The target of Egg Production was 41 billion during 2004-05. Currently, India ranks fifth in egg production in the world.

*Wool Production.* The wool production in the country was 48.5 million kg in 2003-04. Wool Production target for 2004-05 was set at 50 million kg.

*Other Livestock Products.* Livestock sector not only provides essential protein and nutritious human diet through milk, eggs, meat, etc., but also plays an important role in utilisation of non-edible agricultural by-products. Livestock also provides raw material/by products such as hides and skins, blood, bone, fat, etc.

### Cattle and Buffalo Development

India possesses 27 acknowledged indigenous breeds of cattle and seven breeds of buffaloes. Various Central and Centrally sponsored schemes are being implemented for genetic improvement of cattle and buffalo with a view to enhance the per capita availability of consumption of milk through increased milk production. Efforts are also made to protect and preserve the indigenous cattle and buffalo in their native tract, which are facing threat of extinction. The elite animals are selected and registered on the basis of their performance for production of superior pedigree bulls, bull-mothers, frosen semen and frozen embryos for future breeding improvements.

The National Project for Cattle and Buffalo Breeding envisages 100 per cent grant-in-aid to implementing agencies. The Project will also promote about 14,000 private artificial insemination (AI) practitioners and buildup an annual frozen semen production capacity of 66 million doses. Since inception, 26 States have been assisted with Rs 202.52 crore up to 31 March 2005 for participating in the Project.

A Central Herd Registration Scheme for identification and location of superior germ plasm of cattle and buffaloes, propagation of superior germ stock, regulating the sale and purchase, help in formation of breeders' society and to meet requirement of superior bulls in different parts of the country is also being implemented. The Government has established Central Herd Registration Unit in four breeding tracts, i.e., Rohtak, Ahmedabad, Ongole, Ajmer. The total 92 milk-recording centres are functioning to register these breeds of cattle, viz., Gir, Kankrej, Hariana and Ongole and in Buffalo Jasffrabadi, Mehsani, Murrah and Surti.

The seven Central cattle breeding farms at Suratgarh (Rajasthan), Chiplima and Sunabeda (Orissa), Dhamrod (Gujarat), Hessarghatta (Karnataka), Alamadi (Tamil Nadu) and Andeshnagar (Uttar Pradesh) are engaged in scientific breeding programmes of cattle and buffaloes and production of high pedigreed bulls for National Project for Cattle and Buffalo Breeding Programme besides providing training to farmers and breeders. During 2004-05, these farms produced 313 bull calves and supplied 252 high pedigreed bull calves for use under Artificial Insemination Programme in various parts of the country. The Central Frozen Semen Production and Training Institute, Hessarghata (Karnataka) produced 12.83 lakh doses and supplied 11.921akh doses of frozen semen of high pedigreed Sahiwal, Red Sindhi, Holstein Friesian, Jersey, crossbred and Murrah buffalo to different states for their AI programmes.

### Poultry Development

The value of output from poultry sector is nearly 15000 crore and it provides direct or indirect employment to over 2 million people. About 25 per cent of the total egg production in the country comes from desi poultry, which is unorganised rural backyard system. A target for achieving production of over 52 billion eggs by 2011-12, at a growth rate of 4.3 per cent has been visualised by the Government of India.

Poultry sector, besides employment generation and subsidiary income increase, provides nutritional security especially to the rural poor. Further, landless labourers derive more than 50 per cent of their. income from livestock especially poultry.

Central Poultry Development Organisations have been playing a pivotal role in the implementation of the policies of the Government with respect to poultry as a tool for alleviating nutritional hunger and palliating the impecuniosities of the resource-poor farmers especially the women. The mandate of the Central Poultry Development Organisations has been specifically revised, by restructuring all poultry units of this Department to focus on improved indigenous birds, which lay on an average 180-200 eggs per annum and have a vastly improved FCR ratio in terms of feed consumption and weight gain. The Central Poultry Development Organisations have been entrusted with the responsibility of producing excellent germplasm in the form of day old chicks and hatching eggs of these varieties like Nirbheek, Hitkari, Vanaraja, Shyama Cari, CHABRO, etc. Besides, these Organisations are imparting training to approximately 1200-1500 farmers every year both in-situ and ex-situ. These Organisations are also playing a crucial role in analysing feed samples.

These Organisations, besides the activities stated above, scaling-up of diversification of other avian species like Ducks/ Turkeys/ Guinea fowl/ Japanese Quail, upgradation of Training Unit into International Tropical Avian Management Institute in which private-public partnerships is envisaged. Presently these Organisations are also supporting and handholding the Centrally Sponsored Scheme, "Assistance to State Poultry Farms."

The Random Sample Poultry Performance Testing Centre(RSPPTC), Gurgaon is entrusted with the responsibility of testing the performance of layer and broiler varieties. This Centre conducts one layer test for the period from 0-72 weeks of age wherein the production parameters of the laying birds received from different participants comprising of private and public sector agencies are tested. At the end of the test the data are analysed and the report indicating the relative performance of these entrants are published. Accordingly two broiler tests are conducted every year from 0-8 weeks of age and the results are published giving full information about growth rate, feed efficiency and margin of profit from each entrant. The report is published for the utilisation of the poultry industry in general, and the breeder participants in particular. This Centre gives a valuable information relating to the different genetic stock available in the country.

A new centrally sponsored scheme called Assistance to State Poultry Farms is being implemented during the Tenth Plan where one-time assistance is provided to suitably strengthen the farms in terms of hatching, brooding and rearing of the birds with provision for feed mill and their quality monitoring and in-house disease diagnostic facilities. During the first three years of the Tenth Plan, Rs 41.00 crore has been released for assisting more than 70 farms.

A new scheme *"Dairy/Poultry venture Capital Fund"* has been launched during the year 2004-05, wherein there is a provision to grant subsidy on- interest payment. Components like establishment poultry breeding farm with low input technology birds and also for Ducks/ Turkey / Guinea Fowl/Quail/Emu/Ostrich, etc., establishment of feed go--down/ feed mill, feed analytical "laboratory, marketing of poultry products (specialised transport vehicles, cool room storage facilities and retention sheds for birds, etc.), Egg grading, packing and storage for export capacity, Retail poultry dressing unit (300 birds, per day), Egg/broiler carts for sale of poultry products and Central grower unit, etc., are considered for financing under this scheme. The nodal agency for the implementation of this scheme is NABARD through nationalised commercial banks.

### *Sheep Development*

According to Livestock Census 2003/ there are about 61.47 million. sheep and 124.36 million goats in the country. About five million households in the country are engaged in the rearing of small ruminants (sheep, goats and rabbits) and other allied activities. The production of wool was 48.5 million kg during 2003-04. The estimated wool production was about 50.0 million kg during 2004-05.

Central Sheep Breeding Farm, Hissar is producing acclimatised exotic/ cross bred superior quality rams. The farm has supplied 510 rams and 50 ewes during 2003-04 to different states. During 2004-05, the farm has supplied 1233 rams and 50 ewes till February 2005.

### *Conservation of Threatened Breeds*

The population of some of the purebred small ruminants, equines, pigs and pack animals has come down considerably and such breeds have come to the category of threatened breeds in the country. The farms or the farmers w1it in their respective breeding tract are to be established with 100 per cent Central assistance for breeds of these animals wherein their population is less than 10/000 with active participation of State Governments and NGOs, etc.

A new Centrally Sponsored Scheme for conservation of such threatened breeds has been started during Tenth Five Year Plan with a budget outlay of Rs 1500 lakh. During 2004-05, an amount of Rs 527.68 lakh has been released for conservation of Bhutia Pony (Rs 25.00 Lakh), Doom Pig (Rs 44.38 lakh) and Yak (Rs 50.00 lakh) of Arw1achal Pradesh, Hazi Yak of Sikkim (Rs 58.50 lakh), Yak (Rs 37.00 lakh) and Spiti Pony (Rs 31.45 lakh) of Himachal Pradesh, Malabari Goat (Rs 28.55 lakh) and Angamaly Pig (Rs 20.00 lakh) of Kerala, Manipuri Pony (Rs 77.00 lakh) of Manipur, Mithun (Rs

44.50 lakh) of Mizoram, Long Haired Goat of Nagaland (Rs 61.30 lakh) and Ghoongro pig (Rs 22.35 lakh) and Bonpala Sheep (Rs 27.65 lakh) of West Bengal. There is a budget provision of Rs 600 lakh under the scheme during 2005-06.

### *Meat Production/Processing and Export*

A Centrally Sponsored Scheme "Assistance to States for modernisation/ improvement of abattoirs and establishment of CClrcc1SS utilisation centres and primary hide flaying units" was implemented. The objectives of the scheme were to provide wholesome and hygienic meat, gainful utilisation of animal by-products, prevention of environmental pollution and cruelty to animals. Under the scheme, financial assistance of Rs 14.15 crore was provided for incomplete projects to States of Arunachal Pradesh, Maharashtra, Rajasthan, Andhra Pradesh, Tamil Nadu, Uttar Pradesh, Nagaland, Kerala, Tripura and Uriion Territory of Chandigarh during 2004-05. The scheme has now been discontinued as per Zero Based Budgeting (ZBB) decision.

### *Piggery Development*

There are more than 128 lakh pigs in the country of which approximately 14.5 per cent are graded and exotic variety. There are about 158 pig breeding farms in the country run by the State Governments/Union Territories. Exotic breeds like Large White Yorkshire, Hampshire and Landrace are maintained at these farms.

A Centrally-sponsored Scheme namely, Assistance to States for Integrated Piggery Development, was implemented for strengthening the pig breeding farms in the States during Ninth Plan. During the period, an amount of Rs 1,521.87 lakh was released for strengthening the pig breeding farms in various States. The scheme was discontinued at the start of Tenth Plan.

A new scheme on piggery development viz., 'Integrated Piggery Development' has been initiated, on the basis of evaluation of the earlier scheme by NABARD, during 2005-06. Integrated Piggery Development scheme will be part of a new restructured centrally sponsored macro management scheme viz., National Project for Improvement of Poultry and Small Animals.

### *Feed and Fodder Development*

For improvement of vast livestock resources through proper scientific methods, availability of nutritious feed and fodder is essential. To make available scientific fodder production technology, seven regional stations have been established in different agro-climatic zones. These are also engaged in production and propagation of certified quality fodder seeds. During 2004-05, these stations produced 178.00 MT of fodder seeds and conducted 3480 field demonstrations of new fodder varieties.

A central fodder seed production farm at Hessarghatta (Karnataka) produced 67.10 MT seeds of different varieties of fodder grasscs/legumes seeds during 2004-05 and conducted 637 field demonstrations of new fodder varieties. A Central minikit demonstration programme of fodder crops is under implementation for popularising high yielding fodder varieties on a large scale. During 2004-05, 4.90 lakh minikits were allotted to the States for distribution to the farmers free of cost.

Besides, a Centrally Sponsored Scheme, viz., Assistance to State for Feed and Fodder Development is being implemented for establishment of fodder banks and enrichment of straw and cellulosic waste. This scheme was weeded out since 2002-03 but owing to the drought situation in the country, two components of the above scheme, i.e., establishment of fodder banks and enrichment of straw/cellulosic waste were revived up to March 2005 on 75 per cent and 100 per cent Central share basis respectively.

During 2004-05, financial assistance of Rs 245.25 lakh has been provided for enrichment of straw/cellulosic waste and Rs 944.75 lakh provided for establishment of fodder bank.

### *Dairy Development*

The Indian Dairy Industry has acquired substantial growth from the Eighth Plan onwards, achieving an annual output of over 88.1 million tonnes of milk. India's milk output has not only placed the industry first in the world, but also represents sustained growth in the availability of milk and milk products. The Government implemented three schemes in the dairy sector during 2004-05.

*(a) Integrated Dairy Development Projects in Non-Operation Flood, Hilly and Backward Areas.* The scheme, since modified as Intensive Dairy Development Programme on the basis of the recommendation of the evaluation studies, launched during the Eighth Plan period is being continued during the Tenth plan with an outlay of Rs 175.00 crore. So far, 62 projects with an outlay of Rs 334.08 crore have been sanctioned in 25 States and one UT. A sum of Rs 334.71 crore has been released to various State Governments up to 31 Marcr 2005 and 164 districts have been covered. The scheme has benefitted about 7.12 lakh farm families and organised about 11,600 Village Level Dairy Cooperative Societies till 31 March 2005.

*(b) Strengthening Infrastructure for Quality and Clean Milk Production.* A new centrally sponsored scheme has been launched during 2003-04 in October 2003, with the main objective of improving the quality of raw milk produced at the village level in the country. Under this scheme, assistance is provided for training of farmers on good milking practices. The scheme is being implemented on 100 per cent grants-in-aid basis to the State Governments/UTs for components viz., training of farmer members, detergents, stainless steel utensils, strengthening of existing laboratory facilities whereas 75 per cent financial assistance is provided for setting up of milk chilling facilities at village level in the form of bulk milk coolers. The scheme has a budget outlay of Rs

30.00 crore for the Tenth Plan and during 2004-05, an amount of Rs 24.64 crore has been released to various

States. Since inception, 56 projects in 17 States at a total cost of Rs 66.57 crore with a central share of Rs 54.59 crore have been approved upto 31 March 2005 under this scheme.

*(c) Assistance to Cooperative.* The scheme aims at revitalising the sick dairy cooperative unions at the district level and Cooperative federations at the state level. The scheme is being continued during the Tenth Five Year Plan with an outlay of Rs 130 crore. During the year 2004-05, an amount of Rs 9.82 crore was released by the Government for rehabilitation of 14 sick milk unions against budgetary allocation of Rs 15.00 crore. The allocation for the year 2005-06 has been fixed at Rs 10.00 crore for continuation of the scheme.

*(d) Milk and Milk Product Order-1992.* The Government notified the Milk and Milk Product Order on June 1992. As per the provisions of this order, any person/dairy plant handling more than 10,000 litres per day of milk or 500 MT of milk solids per annum needs to be registered with the registering authority appointed by the Central Government.

The Order was amended from time to time as per the decision taken in Milk and Milk Product Advisory Board and as per request received from State Governments. As per the amendment dated 26 March 2002, the provisions of assigning milk shed have been done away with. The power of granting Registration to the units up to two lakh litres per day processing capacity where entire activities of units lies within a state has been delegated to concerned State Registering Authority.

So far, Central Registering Authority and State Registering Authority have granted registration to 751 units with a combined capacity of 841.48 lakh litres per day of milk up to 31 March 2005.

## Fisheries

The Department of Animal Husbandry, Dairying and Fisheries has been undertaking various production, input supply and infrastructure development programmes and welfare-oriented schemes besides formulating/initiating appropriate policies to increase production and productivity in the Fisheries Sector.

The fisheries sector has been one of the major contributors of foreign exchange earnings through export. Export of fish and fishery products has grown manifold over the years. From about 15,700 tonnes valued at Rs 3.92 crore in 1961-62, exports have grown to 4.12 lakh tonnes valued at Rs 6091.95 crore in 2003-04.

**Fish production since 1980-81**

*(lakh tonnes)*

| *Year* | *Marine* | *Inland* | *Total* |
|---|---|---|---|
| 1980-81 | 15.55 | 8.87 | 24.42 |
| 1990-91 | 23.00 | 15.36 | 38.36 |
| 1991-92 | 24.47 | 17.10 | 41.57 |
| 1992-93 | 25.76 | 17.89 | 43.65 |
| 1993-94 | 26.49, | 19.95 | 46.44 |
| 1994-95 | 26.92 | 20.97 | 47.89 |
| 1995-96 | 27.07 | 22.42 | 49.49 |
| 1996-97 | 29.67 | 23.81 | 53.48 |
| 1997-98 | 29.50 | 24.38 | 53.88 |
| 1998-99 | 26.96 | 26.02 | 52.98 |
| 1999-2000 | 28.52 | 28.23 | 56.75 |
| 2000-01 | 28.11 | 28.45 | 56.56 |
| 2001-02 | 28.30 | 31.20 | 59.56 |
| 2002-03(P) | 29.90 | 32.10 | 62.00 |
| 2003-04(P) | 29.41 | 34.58 | 63.99 |
| 2004-05(P) | 30.10 | 35.00 | 65.10 |

P: Provisional

## Development of Inland fisheries and Aquaculture

The ongoing scheme of Development of Freshwater Aquaculture and Integrated Coastal Aquaculture have been combined with four new programmes on Development of Coldwater Fish Culture, Development of Water-logged Area and Derelict water bodies into Aquaculture Estates, Use of Inland Saline / Alkaline Soil for Aquaculture and programme for augmenting the Productivity of Reservoirs. This scheme broadly has two components - Aquaculture and Inland Capture Fisheries.

## Development of Freshwater Aquaculture

The *Government* has been implementing an important programme in inland sector, viz., Development of Freshwater Aquaculture through the Fish Farmers Development Agencies (FFDAs). A network of 429 FFDAs covering all potential districts in the country are in operation. During 2003-04, about 25,394 ha of water area was brought under fish culture and 34,123 fish farmers were trained in *improved* aquaculture practices through FFDAs.

### Development of Brackish Water Aquaculture

With the objective of utilising the country's *vast* brackishwater area for shrimp culture, an area of about 26,365 hectares was developed for shrimp culture till 2003-04 through 39 Brackishwater Fish Farmers Development Agencies (BFDAs) set up in the coastal areas of the country. The agencies have also trained 25720 Fishers in improved practices of shrimp culture till 2003-04. Presently about 50 per cent of the shrimp exported from the country is from aquaculture.

### Development of Marine Fisheries

The Government is providing subsidy to poor fishermen for motorising their traditional craft, which increases the fishing areas and frequency of operation with consequent increase in catch and earnings of fishermen. About 42,050 traditional crafts have been motorised so far. The Government has also been operating a scheme on fishermen development rebate on HSD oil used by fishing vessels below 20 metre length to offset the operational cost, incurred by small mechanised fishing boat operators.

### Development of Fishing Harbour

The *Government* has been implementing a scheme with the objective of providing infrastructure facilities for safe landing and berthing to the fishing vessels. Since inception of the scheme, six major fishing harbours *viz.*, Cochin, Chennai, Visakhapatnam, Roychowk, Paradip and Sassoon dock (Mumbai), 38 minor fishing harbours and 142 fish landing centres have been constructed. 17 minor fishing harbours and 46 fish landing centres are at various stages of construction.

### Welfare Programmes for Traditional Fishermen

Important programmes for the welfare of traditional fishermen are :

(i) Croup Accident Insurance Scheme for active fishermen;

(ii) Development of Model Fishermen Villages; and

(iii) Saving-cum-Relief Scheme.

The fishermen icenlified or registered with the States/UTs are insured for Rs 50,000 agains'i death or permanent total disability and Rs 25,000 against partial disability. About 14.74 lakh fishermen were insured during 2004-2005. Under the Development of Model Fishermen Villages, 12,458 houses have been sanctioned for the benefit of fishermen in 2004-2005. Under the Saving-cum-Relief component, financial assistance is provided to the fishermen during the lean fishing season. About 2.63 lakh fishermen were assisted under the saving-cum-relief programme in 2004-2005.

## Specialised Institutes

The Central Institute of Fisheries, Nautical and Engineering Training, Kochi, with units at Chennai and Visakhapatnam, aims at making available sufficient number of operators of deep-sea fishing vessels and technicians for shore establishments. Integrated Fisheries Project, Kochi, envisages processing, popularising and test marketing of unconventional varieties of fish. The Central Institute of Coastal Engineering for Fisheries, Bangalore, is engaged in techno-economic feasibility study for location of fishing harbour sites. Fishery Survey of India (FSI) is the nodal organisation responsible for survey and assessment of fishery resources under the Indian Exclusive Economic Zone (EEZ).

## Agricultural Research Education

The department of Agricultural Research and Education (DARE) Ministry of Agriculture is responsible for coordinating research' and educational activities in agriculture, animal husbandry and fisheries. Besides, it helps to bring about inter-departmental and inter-institutional collaboration with national and international agencies engaged in the same and allied fields. Department provides government support, service and linkage to" the Indian council of Agricultural Research (ICAR).

## Indian Council of Agriculture Research

Alleviating hunger, providing employment and increasing farm income is the major mandate of the agriculture sector. During the year 2004-05, the council generated farm-worthy technologies to further enhance agricultural growth in the country and ensure national food and nutritional security with enhanced income opportunity. A brief account of the achievements made in agricultural research, education" and extension during the year and initiatives taken by the council towards making India a developed nation through enhanced agricultural productivity and profitability, is presented here.

The Council faced challenging situation due to natural vagaries like floods, late arrival of rains, and the most devastating Tsunami tidal waves on 26 December 2004. In the Tsunami disaster, thousands of precious human lives were lost and property. worth several crores of rupees was destroyed, causing ~thinkable misery to the people. The council spared no efforts in helping the cause of Tsunami victims hand-in-hand with other Government Departments.

The National Agricultrual Science Museum at the National Agricultural Science Centre complex, was dedicated to the nation in November 2004 by the Hon'ble President of India. It is a unique facility in the country that depicts the saga of Indian agriculture. The museum is a storehouse of information and presents chronologically the origin and evolution of Indian agriculture over the time with futuristic outlook.

## Crop Science

As the collection, conservation and optimal use of plant genetic resources is a key requirement for higher agricultural production on a sustainable basis, 7,252 accessions of crops and their wild relatives were collected through 117 explorations. Besides, 19,921 accessions of diverse crops and 41,611 samples of international trials were introduced from various cOlliltries. During the year, the National seed Gene Bank has been enriched with 24,690 accessions. A total of 77,002 samples including 59 transgenics were processed for quarantine clearance and 192 phytosanitary certificates were issued for germplasm export. More than 620 varieties/landraces of 27 crops have been fingerprinted a checklist for potential quarantine pests of cereals has been prepared for India.

Development and release of improved crop varieties has further contributed towards sustained and increased production of field crops. In food crops, 15 varieties of rice, four varieties of wheat and one variety of barley have been released for commercial cultivation in different agro-ecologies. Thirteen cultivars of maize, six cultivars,of sorghum and four hybrids of pearl millet have been recommended for release. In forage crops, two varieties of oat and one variety each of Coax, Setaria and Guinea-grass have been identified for release.

In oilseeds, one variety of *gobhi sarson* and three varieties of niger have been released for cultivation. In addition, one variety of soybeans, three varieties each of groundnut and mustard, nine hybrids of sunflower and one hybrid of castor have been identified for release. One hybrid of pigeonpea has been released for commercial cultivation. Besides, two varieties of pigeonpea, three varieties of mungbean, and one variety each of chickpea, lentil, pea and *rajmah,* have been identified for release in pulse· crops. In arid legumes, one variety each of mothbean, cowpea and horsegram has been identified for release. In commercial crops, two hybrids and four varieties of cotton have been notified for commercial cultivation.

One variety each of sugarcane, toss a jute, white jute, sunnhemp and mesta has been identified for release. Transgenics with *Cry* genes were developed for insect pest resistance of rice, cotton and pigeonpea.

A number of bread and durum wheat genotypes identified were found rich in protein, B-carotene, iron and zinc. For the first time, a nested PCR assay has been developed for molecular detection of grassy shoot disease phytoplasma of sugarcane. Other major achievements in crops include: development of cost-effective eco-and user-friendly technology for ramie, isolation of a new virulent and heat-tolerant strain of *Steinernema masoodi* for management of chickpea pod-borer, preparation of interactive identification key to 41 important families of parasitoids and 25 families of insect-predators, sequencing of B-tubulin gene from *verticillium chalamydosporium,* and the reduction of 20-23 per cent *Helicoverpa armigera* larvae on chickpea by eight insectivorous bird species.

## Horticultural Crops

The horticulture sector has established its credibility in improving productivity of land, generating employment, improving economic conditions of farmers and entrepreneurs, enhancing export potential, and above all, providing nutritional security. Wild and cultivated accessions of horticultural crops were collected and added to the respective gene banks. In mango, one of the richest collections is being maintained at the National Active Germplasm Site in Lucknow. Use of biodynamic pesticides minimised the disease and pest incidence. In citrus, nutrient diagnostics in relation to production levels were developed. Integrated pest management (IPM) modules effectively controlled the blackfly and psylla population in citrus.

In banana, unique accessions *(Musa nagensium, M. itinerans and M. aurantiaca),* two new natural intersectional diploids, a new natural mutant of M. velutina and an intermediary hybrid have been collected Identifies. A group of bio-agents and botanicals was developed for management of crown root disease in banana. Technologies of production of banana juice, fig, pickles, 'jam, sauce and health drinks were transferred to entrepreneurs through training. Diagnostic kits for virus diseases of citrus, potato and banana were also developed. In sapota, softwood grafting during May-June was found to be a very effective method of propagation and is being followed by farmers in north Karnataka. Gynodioecious collections of papaya were superior in fruit quality characters and dioecious ones in yield characters.

Two selections of ber, two cultivars each of pomegranate and fig, and one variety of custard apple have been recommended for commercial cultivation, CIAH Hybrid 1 of ber has been developed and multiplied by patch budding. Halawy, Barkee, Khuneizi and zaidi cultivars of date palm were found suitable for fresh consumption, while medjool was suitable for dry dates consumption.

In vegetables and tubers, Pusa Samridhi, a selection of bottle gourd, and arka Anupama variety of spinach have been released for cultivation, Besides, 22 varieties/hybrids (six each in tomato and brinjal, three in chilli, two in cauliflower and one each in bottle gourd, bitter gourd, Dolichos bean, cowpea and cucumber) have been identified for release in different parts of the country.

A heat-tolerant hybrid HT/92-621 of potato suitable for processing into French fries and chips has been recommended for cultivation in peninsular, coastal parts and for early planting in northwestern plains. In sweet potato, a selection 90/170 has been recommended for Orissa and a high carotene hybrid 80/168 has been proposed for release.

In-vitro mushroom fruiting was successfully achieved in two strains of ***Pleurotus*** spp. using amended wheat 'straw and saw dust. Cryopreservation protocol was developed and standardised for mycelia stalk cultures of 11 edible mushroom strains of six genera. The developed lyophilisation protocol for the preservation of mushroom myceliacultures proved better than the earlier methods used by the International Mycological Institute, England. true-to-the-types of medicinal mushroom, Reishi ***(Ganoderma lucidum)***, were evaluated for fruiting and yield in summer and winter seasons.

Tissue culture protocols were standardised for plantlet production of coconut from plumular tissues and of arecanut from inflorescence tissues. In northeastern region, vegetaqble crops performed well as intercrops in coconut and arecanut gardens. A de-shelling machine was found very effective in reducing drying time and improving copra quality. Four varieties of black pepper and two varieties of turmeric have been proposed for release. More than 75,000 cashew grafts have been distributed to farmers. Intercropping of isabgol with ***kalongi*** in 6:2 ratio proved to be most. economically viable system in Malwa plateau. *Two* promising lines of safed musli were identified.

In field crops + teak + papaya system, the net income was three times higher than sole field crop at dharwas. Bench grafting in aonla was highly successful (90 per cent) and. will ensure availability of ready-to-plant grafts well in time. Knowing the benefits of coconut intercropping, farmers have already started the cultivation of grasses, shade-loving vegetables, and rice seedling nurseries in interspace of coconut crop. Simple techniques were developed for manual extraction of custard apple pulp for Simple techniques were developed for manual extraction of custard apple pulp for Simple techniques were developed for manual extraction of custard apple pulp for preparing ice-cream and for separation of aonla by steaming the fruit in a pressure-cooker.

## Natural Resource Management

In research efforts under natural resource management, soil maps of 12 districts in seven states were prepared for laternate land-use options and crop zoning, and the soil-erosion status map of Assam and Tripura was prepared for undertaking appropriate soil-conservation measures. Installation of mechanical and vegetative measures rehabilitated 48 per cent of wasteland to cultivation. Conservation measures were also developed for rehabilitation of degraded hillocks in the Eastern Ghats.

In different rainfed oilseed-based cropping systems, integrated nutrient management (INM) involving use of various sources of organic manures was, effective in increasing yield. An integrated soil-water-atmosphere-plant model was developed for understanding irrigation requirement of wheat. The INM approach considerabley reduced the nitrogen requirement of rainfed sorghum and greengram.

Research on soil salinity in coastal ecosystem led to sustenance of rice and wheat yields in reclaimed alkali soils, identification of causes and remedial measures for re-sodification of reclaimed sodie soils, possible use of urban and industrial effuents in agriculture, and improvement in water productivity in salt-affected irrigated lands with laser land levelling. Besides, salt-tolerant varieties of rice and mustard have been released/dentifies. Sapota and guava possessed more tolerance than other fruits in the coastal saline soil.

Under alternate land-use systems, agri-horti and agroforestry options were developed for rainy-season sorghum as the area under this crop is declining due to erratic rainfall. A ,Rural Bio-resource Centre was set up in a village of Nalgonda district, for mass propagation of the elite teak plants through multiplication in a shade house, Aerobic rice-based cropping system was found more productive than transplanted rice-based cropping system. Green-manuring with identified genotypes of Sesbania and ***Crotalaria*** increased the yield of rice-wheat system by 24 per cent. In rice-wheat system, diversification option were worked out for improving productivity and resource-use efficiency to mitigate its second-generation problems in Indo-Gangetic plains. On-line database was developed for crops and cropping systems.

The control of obnoxious weed Parthenium with Mexican beetle led to restoration of native vegetation at several' places in Madhya Pradesh. Single application of Fenoxaprop was highly effective in controlling barnyard grass. The interventions of Central Arid Zone Research Institute helped

Rajasthan State Mineral Development Corporation in rehabilitating a wasteland, thereby increasing the greenery, control erosion and providing means for livelihood support to the local people.

## Animal Science

As livestock sector is experiencing fast growth, its resources across regions and species have to be utilised optimally to achieve the goals of efficiency, equity, and sustainability. The Country Report on the status of Animal Genetic Resources of India was prepared. It provides an assessment of the country's animal genetic resources and strategy for their optimum management. Introduction of Booroola facundity gene in Malpura sheep through crossbreeding increased its prolificacy. Dahlem Red bird proved as an ideal resource population for exploring genes related to immunity and growth. Ex-situ conservation was undertaken for Nagori and Rathi cattle.

Milk production of more than 4,500 kg was recorded in first lactation in frieswal cows. In Murrah buffaloes, milk yield was 4,114 kg in 305 days or less. Black Bengal goats were found highly prolific, as multiple births were quite high (73.78 per cent). Baseline survey on pig production status was completed in eastern states. A broiler cross from PAU, Ludhiana, could attain 2,005 gm body weight at 7 weeks of age. Dwarf gene was introgressed into naked neck line to realise effects of both genes in broiler.

Antigenic and molecular make-up completed in the repository of 1,171 isolates of foot-and-mouth disease (FMD) virus. Softwares were developed for forecasting animal deseases based on weather. Vaccine developed against ***Pasteurella multocida*** serotypes B : 2 and A : 1 provided protection for six months. A potent vaccine was developed to protect sheep against bluetongue disease. Bioclimatographs were prepared to depict interrelationship between weather and prevalence of various gastrointestinal parasites. Worn-control programme was developed to minimise the frequency of anthelmintic drenching in different zones. Techniques were developed for the detection of bovine diarrhoea virus *(**Haemonchus contortlls**)* and gastrointestinal parasites. Involvement of fowl adenovirus-12 in causation of litchi desease (IBH-HPS) is the first report in the country. Development of patentable inventions included thermostable immunoprophylactic agent against IBD, vaccine against egg drop syndrome-76, method for diagnostic kit for forecasting equine herpes virus-I, and COFFB-Kit for diagnosis of ***Babesia equi*** infection.

Supplementation of vegetable residue (up to 60 per cent) of high quality roughae did not show adverse effects in cattle. Addition of mustard or groundnut, oil could reduce methane production in cattle. Animals when fed chaffed restricted rice straw. Economical rations and complete feeds

resulted in better digestive efficiency in buffaloes. A low-cost feed-pelleting machine was developed. A sheep flock, maintained on degraded rangeland, when fed on tree leaves during the lean period, sustained their performance. Cost of egg production was found less in birds when these were fed on pearl millet diet. Reduction in the intensity of leg abnormality in birds was observed on giving diet containing cholecalciferol or edible-stemmed vine *(**Cissus quadrangularis**)* plant-meal. Sunflower and soybean oil could counter the adverse effect of aflatoxin in the diet of birds.

Testicular and semen evaluation proved effective in identification of potential breeding bulls. Heat-stress eliminating methods were effective in improving the semen production in cattle. First goat kid was born thr0ugh in-vitro fertilised embryo. Twenty-four transferable embryos were flushed out from single Carole sheep, the highest figure from any sheep breed in tropical environment. In birds, induction of concurrent maturation of oviduct improved the egg production. Immuno-modulation of prolactin hormone could increase the egg production in birds. A rapid coliform test was developed for the detection of coliforms in milk and milk products. Methods were standardised for preparation of products from milk of camel, yak and mithun. Economical value-added products were developed from spent hen, chicken gizzard, and pork skin and were found acceptable to customer. Optimum, concentration of organic acids was determined to reduce microbial concentration in meat, as some countries allow use of only, these acids. .

## Fishery Science

Fisheries sector stimulates growth of a number of subsidiary industries and provides a vital source of low-priced and nutritious foods, besides earning huge foreign exchange and providing nutritional security to the country. Oil sardine recorded an all-time annual landing with an increase of 9.6 per cent. Under inland sector, 33.4 per cent increased fish yield was recorded from the Hooghly-Matlah estuarine system because of higher yield of hilsa. In culture fisheries, important achievements have been the successful cryopreservation of spermatozoa from dead carps, development of eco--friendly cage ·culture for rearing fry and fingerlings in reservoirs, and development of kits for diagnosis of bacterial diseases in freshwater fishes. A product, CIFACURE, was also developed for controlling bacterial and fungal infections in ornamental fishes. In cold-water fisheries, a well--balanced diet was formulated for juveniles of golden mahaseer. Ecological status of high mountain Himalayan lakes was studied to promote fish rearing.

Research efforts in brackishwater aquaculture led to the development of a technique for captive breeding of kuruma shrimp for providing quality broodstock, evolvement of immunodiagnostics for detection of fish pathogens, and development of a diagnostic kit for ***Macrobrachium roseenbergii nodavirus***. The work in mariculture resulted in: successful breeding of doctor shrimp, breakthough in captive breeding and larvae production in sand lobster for the first time in country, and production of good-quality image pearls from flat oyster (hitherto considered a pest). In fish genetics, observation of significant increase in the DNA damage frequency when exposed with geniotoxic pollutants, development of DNA damage frequency when exposed with genotoxic pollutants, development of DNA probe and hybridisation techniques for detection, of shellfish pathogens, and establishment of live gene banks for threatened fish speices are important achievements. In post-harvest technology, for quality assurance of processed freshwater scampi ***(M. rosenbergii)*** and package of practices for value-added ready-to-serve products were developed. In Island development programme, results indicated a good scope for gill nets on the agatti Island in Lakshadweep.

## Agricultural Engineering

In agricultural engineering and technology, laser-guided leveller and banana sucker were the implements developed under tractor-operated machinery. Chipper shedder and lightweight power tiller chipper were fabricated as power tiller-operated machinery. Eight-row pre-germinated rice seeder and engine-operated sugarcane leaf stripperwere developed under self-propelled and stationary machinery respectively.

Calilbration of tractor-operated PAU two-row vegetable transplanter and prototype feasibility testing of MPKV (Rahuri) planter and pneumatic planter were done. Feasibility testing of: PAU walking sprayer and transplanter under the self-propelled machinery, CRIDA bullock-drawn groundnut-planter and CIAE animal-drawn inclined plate planter as the animal-drawn machinery, and power weeder under stationery machinery was also carried out satisfactorily, A pedal-operated potato slicer was developed. As less-skilled people in rural and semi-urban areas use the CIAE developed dal mill, its design was refined. Candies and bars ware prepared from mango, guava, banana and pineapple with different combinations of soymilk, soy-flour, cashewnut, grape-raisin and groundnut kernel, and their production at pilot-plant scale Was found economical. New puffed potato cubes, step-wise expanding pitch fruit grader, starter to be used for hydration of soybean and sunflower seeds, colour meter to measure mango maturity, custard-apple pulper and a new technique of processing mustard seeds were also developed.

A biopesticide ***Bacillus thuringiensis*** ssp. Kurstaki suppressed 71-86 per cent ***Eublemma ambilis*** and 78-89 per cent ***Pseudohypatopa pulverea***, the major predators of lac. A blend ratio of 50:50 sisal : jute could be

processed for making shiny and lustrouswrappers and upholsteries with very high aesthetic value. Scrubber, another product of the blend, proved eco-friendly and can replace synthetic counterparts. A laboratory-scale high-rate biomethane technology was developed for treatment of effluents of sago-industries. A light-sensitive solar tracker was developed to orient solar devices towards the Sun to maximise solar energy harvest. A solar photovoltaic winnower-cum-dryer was fabricated for efficient wiru1.owing of grains, spices and other agricultural produce, and for dehydrating fruits and vegetables. Plastic mulching machine, castor sheller, groundnut stripper and vegetable preservator were also developed. Twenty-eight feasible technologies/ designs were facilitated for adoption by farmers and industries.

## Agriculture Education

For agricultural human resource development, the ICAR provided financial support to the State Agricultural Universities (SAUs): for the development of undergraduate and post-graduate programmes to expand and improve quality and utility in agricultural education and training. The ICAR Accreditation Board granted accreditation to two deemed-to-be Universities (DUs) and eight SAUs to assure and sustain high quality agricultural education. During the year, students from 24 foreign countries were admitted in various degree programmes in ICAR-DUs/SAUs. In Summer/ Winter Schools/Short Courses and Centres of Advanced Studies (CAS), 2,925 scientists or faculty members were trained in diverse disciplines of agriculture and allied sciences. Eighteen scientists from ICAR Institutes/SAUs were deputed abroad for developing core competence in biotechnology. The National Talent Scholarship was awarded to 230 candidates, and Junior Research Fellowship (JRF) was given to 456 candidates. The ICAR has enhanced the scholarship at Master's level, Ph.D. level and for students of the four DUs. Guidelines have been rationalised for National Professor Scheme.

The National Academy of Agricultural Research Management (NAARM) trained 1,354 officials from the ICAR, and other institutions and organisations, besides conducting two international training programmes under ICAR-CARP (Sri Lanka) Plan.

In the area of social sciences and policies, allocation of scientific manpower in ICAR was analysed for better manpower planning. Dairy contract farming was found to reduce the transaction cost to the participants. Studies on activity based typology of mixed crop-livestock system were undertaken to improve the livestock productivity. Adoption of zero tillage technology in wheat was found to save the cost by 28 per cent.

## Agricultural Extension

Now, there are 440 Krishi Vigyan Kendras (KVKs), 76 Technology Assessmentand Refinement through Institution-Village Linkage Programme (TARIVLP) centres and 44 Agricultural Technology and Information Centres (AIICs). Through these institutional mechanisms, a number of steps were taken to bridge the gap between the technology developed at the reseearch institutions and its adoption at the field. The ATICs provided 736.3 tonnes seed of improved varieties, 0.79 million planting materials, 4.03 million packets of biofertilizers and biopesticides and, 15,405 implements farmers. Besides, the processed products to end printed and electronic information material to farmers and stakehoders and other users. KVKs organised 30,517 training programmes, benefiting 610,000 farmer and farmwomen 130,000 rural young and 68,708 in-service personnel. For acceleration of the dissemination process, Kisan melas, discussion fora, advisory services and diagnostic film shows were arranged.

The KVKs also took up on-farm testing of 607 farm technologies. A total of 32,340 frontline demonstrations, covering production technologies on oilseeds, pulses and other crops, besides carrying out 3,327 demonstrations on livestock production and management. At KVKs, 7,490.73 tonne seed of cereals, oilseeds, pulses and vegetables, spices, medicinal and aromatic plants, ornamental and plantation crops, and forest species; and 958,670 livestock strains were produced and made available to the farmers. Significant impact of KVK technological interventions is also visible in villages in terms of rainwater harvesting, empowerment of rural and tribal women through income generation activities, dairying as sustainable enterprise for women, reclamation of sodic land, off-season vegetable poduction, backyard poultry and revival of ginger cultivation in different states.

Technology interven,tions (2,219) through rVLP were undertaken in selected production systems under five agro-ecosystems with major focus on NRM, IPM, biodiversity, IPNM, PHT and value-addition, water management and socio-economics. Under Mission Mode Project on Indigenous Technical Knowledge (ITK), geographical indicators of the trees of plants, which are in use in ITKs, have been documented. Now 4033 documented ITKs are available.

The NRC for Women in Agriculture is addressing gender issues in agriculture through adoption of various approaches to engendering agricultural research and extension and projects like development of farming system suitable for women in. agriculture. Technological and economic empowerment of women in agriculture, compilation of 24 hrmen-women work profile, introduction of drudgery-reducing implements, establishment of 40 self-help groups and building food security through rural agriculture are important achievements. The self-help group were able to raise 150,000 saplings of bio-diesel plant, Jatropha. The women fish seed-groups of Orissa received 2.4 million ***Rohu*** and 0.8 million grass carp

seed. Model nutrition gardens were established in adopted villages to motivate the community to replicate these in their own fields to achieve cost-effective solutions for micronutrient deficiencies.

## Research for Tribal and Hilly Areas

Major emphasis was given on agricultural research for the benefit of tribal and hilly regions. Four varieties (one each of wheat, rice, finger millet and garlic) were released from the Vivekananda Parvatiya Krishi Anusandhan Shala, Almora, for nothwestern hills. Twenty-seven genotypes of wheat, resistant to different diseases, were also identified. Hybrid Napier proved to be the best grass for fodder production as well as for terrace stabilisation **in** hills. A cost-effective and user-friendly light trap was developed for white-grub beetles. A lightweight seed-cum-fertilizer drill for the sowing of wheat, barley, rice and pulses was developed for hilly soils. Kodomillet--soybean rotation was found acceptable in tribal areas of Madhya Pradesh.

Research conducted at the Central Agricultural Research Institute, Port Blair, recommended cultivation of a very early rice variety ***Heera*** followed by medium-duration varieties in double cropping system. This could replace the low-yielding ricevariety C 14-8. Mapping population of **IR** 29 x Pokkali to tag salt-tolerant genes, and the aluminium and iron-toxicity tolerant rice using somaclonal variation were developed. A compendium on biodiversity of mangrove ecosystem of Andamans was brought out. Important findings in livestock are: suitability of III 80 (M) x Brown Nicobari (F) progeny for .backyard farming, standardisation of housing and management conditions for Japanese quails in islands, higher adult body weight of chicken (M) x quail (F) crosses for commercial purpose, and standardisation and popularisation of poultry products and post-harvest technologies.

At the ICAR Research Complex for NEH Region, Umiam, the developed agri-horti-silvi-pastoral model followed by agriculture-based land-use system COl;lld reduce soil loss with in-situ water retention. Sloping agricultural land-use technology seemed suitable for restoration of ***jhum*** fallow or degraded land in eastern Himalayas. Identification of disease- and moisture stress-tolerant tomato, development of immunodiagnostic kit for rapid diagnosis of gastro-intestinal parasites from infected animals and standardisation of PCR-based diagnostic protocols for ***Clostridium, Pasteurella*** and ***Salmonella*** are some other major achievements.

National Agricultural Technology Project

The World Bank Final Review Mission has adjudged the National Agricultural Technology Project (NATP) as highly rewarding. The project is considered to have achieved its overall objective of increasing the availability and

adoption of appropriate technologies, besides being found effective in initiating reforms in our agricultural research to make it more efficient and responsive to agricultural technological needs of the country.

In rainfed agro-ecosystem programme under NATP, the water stored in refuges was utilised to grow the winter crops and raise fruit plants on embankment .of the refuges. Vegetative barriers like stylo and hill broom and suitable tree species like guava, mulberry and ***subabul*** for bund plantation resulted in the control of soil erosion and runoff in Orissa, Andhra Pradesh and Chhattisgarh thus leading to rehabilitation of degraded hillocks. A low-cost push-type safflower harvester was developed. In irrigated agro-ecosystem under NATP, three varieties of under-exploited crop species, viz., two green-manuring varieties (Pant Ses. 1 and DH 1) of Sesbania and one variety (Narendra Sanai 1) of Crotalaria, were released.

In coastal agro-ecosystem under NATP, antiserum was developed against Ganoderma causing basal rot of plantation crops and a protocol standardised for detection of its infection in situ. The cultivation of oyster mushroom could be popularised through women self-help group. Under NATP, a process for saffron pigment and flavour concentrate extraction was developed for utilisation in pharmaceutical and food industries and the protocol for high frequency protoplast isolation and regeneration in Agaricus species was standardised. Antibody quantification kits were prepared and validated for infectious bursal disease, infectious bronchities and egg drop syndrome virus. Now, four Referral Laboratories for quality assurance of crops, horticultural, fish and livestock products, and cotton textile for export and domestic markets are available. A DNA vaccine was developed against brucellosis in goat. For the first time in the world, a competitive inhibition ELISA was standardised using IgM monoclonal antibodies (mab) c(clone 9E6) against recombinant gag antigen of bovine immuno deficiency virus. In a survey of northeastern states, 250 fish species have shown potential as ormamental species. An integrated farming model was found to generate additional income of Rs 19,700/ha/year in acid saline wetlands of Kerala.

The Council has filed 52 applications for patents of new technologies, and assigned 31 technologies to the National Research and Development Corporation for commercialisation. Fifty-two awards were bestowed in 12 different categories to honour

42 scientists and their associates, four institutions, two All-India Co-ordinated Research .projects, three farmers and one agricultural journalist. The Council has also enhanced the prize money from Rs 100,000 to 500,000 for Sardar Patel Outstanding Institution Award; and from Rs 20,000 to 50,000 for Bharat Ratna Dr. C. Subramaniam Award for outstanding teacher. Financial assistance was provided to 46 scientific societies for bringing out journals, and 27 societies and academic universities for conducting semuldrs/symposia/conferences.

The ICAR has taken several initiatives under organisation and management reforms to improve working environment to make research effective, efficient and relevant. The Council settled 929 audit paras out of 1,353 as of May 2004 in respect of all the ICAR Institutes and generated internal resources amounting to Rs 255.6 million during 2003-04 as against Rs 214.3 million during the previous year; and fixed a target of Rs 5695 million for 2004-05. The Asset Registers in respect of 103 accounting units out 111 were updated. Personnel Management Information System Network was developed for systematic planning and better development of human resources of the ICAR. Awareness Creation Unit was established at ICAR headquarters.

## Partnership and Linkages

The DARE and ICAR have been operating partnership and linkages at national and international level through ·Memoranda of Understanding (MoUs)/Work Plans, Training Courses, Exchange Visits, etc. Three MoUs were signed between the Government of India and Kingdom of Morocco; the ICAR and International Livestock Reasearch Institute; and the National Centre for Agricultural Economics and Policy Reasearch and Asian Development Bank Five Work Plans were signed by ICAR with the Agricultural Reasearch Centre Egypt; International Centre for Tropical Agriculture, International Plant Genetic Resources Institute, International Water Management Institute, and the World Fish Centre for scientific and technical co-operation in the field of agriculture and education. Under international linkages, five projects were approved/intiated. Delegations led by the President of Tajik Academy of Agricultural Sciences, Tajikistan; Agriculture Minister of Bhutan; Governor of Henan, China; and Director-General of Ethiopian Agricultural Reasearch Organisation, Ethiopia besides others visited India. Indian scientists/technicians visited foreign coutries to understand their recent advances in agriculture and allied subjects. An Indo-Iran joint workshop was held on Drought Management.

The Directorate of Information and Publications of Agriculture (DIPA) brought out 48 publications in English and 15 in Hindi. The Agricultural Research Information Centre developed database covering the biblographical details of all research information published in India in the field of agriculture and allied sciences. The DIP A has procured information on 3,500 inputs and added this to the National Agricultural Research Database.

# 3

# Ancient History of Agriculture

Evidently, the history of agriculture in India begins with the Indus Valley riparian culture. The alluvial soil, heavy monsoon rainfall and the water from the Indus, provided substantial agricultural support to the teeming populations of Mohenjo-Daro and Harappa. Besides, The river made transport, trade and irrigation easy. One the ideograms reveals the use of a toothed harrow for cultivation. Flood-irrigation and the arrangements for drawing water for irrigation by the use of a certain jar type pottery and long rope on contrivances like the wheel, similar to that utilised in the Near and Middle East, have been mentioned by Marshall and Mackay.

The wheat unearthed during the Mohenjo-Daro period, which were identified as *Triticum sphaerococcum and T. Compactum,* are still cultivated in the region of modern Punjab. Evidences exist that agriculture during those days comprised cultivation of wheat, barley, sesamum, peas, cotton date palm, pomegranate and banana.

The querns (a stone handmill) found in the excavations sow how the grinding of the grain was done on an extensive scale. Invocations to Gods of plenty and prosperity in Rig Vedic and Atharvavedic hymns indicate that agriculture was an important vocation during the Vedic period. Agricultural practices had social and religious significance. Domestic rites and festivals during the Vedic period synchronized with the four main agricultural operations like plougning, sowing, harvesting and threshing, animal husbandry and maintenance of cattle. This practice is continued until this day.

Similarities existing in the agricultural expressions in Vedic India and Iran suggest possible similarities in the agricultural practices undertaken, at that time, in these two countries.

The mention of a plough *(langala)* drawn by six, eight, or twelve oxen to bring the soil to the desired tilth by repeated ploughing; sieve *(titau)* to separate grain from the straw, vessel *(urdara)* to measure grain, choice of the use of dried cow-dung *(karisa,* as opposed to fresh cow-dung), sakrt, as manure in several references of the Rig Veda and the process of cultivation and harvesting using the plough and sickle, respectively in *Kathaka Samhita,* all point to the level of agriculture during this period.

Agricultural production for the Vedic people had the same connotation as human reproduction, for mention has been made of cultivation of crops in *Atharvaveda* to "Scatter the seeds in the prepared ground (yoni)."

Even though mention exists of the cultivation of wheat and barley during the Rig Vedic period, rice must have been growing wild. It was not until the time the *Taittiriya, Kathaka Samhitas* and the *Atharvaveda* were written that rice must have been domesticated and the technology developed because references to good quality rice *tandula, vrihi, sali*) as opposed to wild variety (*nivara*), husked (karna) and unhusked (akarna), and rotation of crops, involving sowing of rice in summer and pulses in the winter, exists. During this period two harvests were undertaken. Roxburgh,the famous botanist, acknowledges crop rotation as an Indian contribution to the Western world. The farmer seemed to possess a fairly good knowledge of the fertility of the eland, selection and treatment of the seeds, seasons of sowing and harvesting, rotation and other practices, manuring for increased production etc.

Amongst the several domestic animals were bulls, oxen, sheep, goat and cows. The usefulness and importance of the cow is extolled in the Rig Vedic hymns.

During the Vedic period farmers were also conversant with the principles of collecting water from catchment areas in hilly terrain and carrying it through canals. Mention also exists of *'asmachakra'*, a water lifting wheel device made of stone to which pails were strapped with leather belts, *'drona'*, *'ghatayantra'* or *udghatana'* which are drum shaped wheels to which '*ghata*' or earthen pots were tied at equal distances which were operated by endless ropes.

During the post-Vedic, Magadha period, when political unification was attempted, taxes were collected from subjects who followed different livelihoods, vocations like agriculture *(krishi),* cattle raising *(pasupalana)* and trade *(vanijya).*

The *Arthasastra* of Kautilya also mentions settlement of subjects in unoccupied *(sunyaivesa)* lands with an aim to increase agricultural production. Construction and maintenance of irrigation works ensured continued productivity of land. Keeping fallow of fertile uncultivated land was considered a culpable offence. Land use classification based on the status of cultivation or position was well documented.

The *Arthasastra* also defined the qualifications and job descriptions of the "*Sitadhyaksa*" or the Director of agriculture, who should be conversant with the suitability of the soil, ploughting, preparation of the soil, selection of seeds etc., and correlate to the rainfall during the different seasons of the year.Panini's land use classification included cultivated (*karsa*) and uncultivated waste land (*usara*) and pasture (*gochara*); crops grown and the quantities of seed required for sowing. Three crops, sown respectively in the rainy season and stored in Magha (January-February); in autumn and stored in Chaitra (March-April); and third sown in spring stored in Jyestha (May-June) were mentioned during this period. Knowledge of the requirements of different crops led to a rudimentary form of crop rotation. the resources and sometimes the state intervened in the additional crops that should be undertaken by the farmers in the interest of taxes that can be collected. Patanjali also informs of the mixed crops that were also undertaken.

Megasthenes records that the Mauryan monarchs considered construction of tanks and reservoirs for irrigation as acts of welfare and religious merit, which the State was expected to provide. Thus, Pushyagupta during the reign of Chandragupta Maurya constructed the Sudarshana Lake in Saurashtra, which was later restored by Tusaspa, the Yavana Governor of Ashoka. The breach sufferend by this reservoir was repaired by the Parthian Minister (*amathya*) Suvishaka, the Governor of Rudradaman, the Saka ruler of that time. the successors of Rudradaman, the Saka ruler of that time. The successors of Rudradaman continued this tradition in western India.

The central and south Indian tanks, constructed with great engineering skill during the beginning of this civilisation, still serve the purpose for which they were made. The embankments on the river Kaveri, and the several canals, constructed by Karikala I during the first century A.D., for utilisation of the water bear testimony to this effort. During this period, tank sub-committees of the village assemblies maintained the upkeep of the tanks.

Under royal patronage like the Guptas in the Gangetic basin, Malwa, Gujarat and Kathiawar; of the Vakatakas and Chalukyas in the deccan; and the Pallavas in the south agriculture got diversified and agricultural product-based trade flourished in different parts of this country during the Classical Age and later.

The Chinese pilgrim, Hiuen Tsang, who visited India, while describing the agricultural prosperity of India, during that period, has recorded the existence of a short duration rice (maturing in 90 days), wheat, ginger, mustard, tamarind, mango, melon, wood-apple, pear, plum, peach, apricot etc.

Encyclopaedic references of the period like *Amarakosha* and the *Brthatsamhita* have special chapters for agriculture called Krishi Parasara which include the existing knowledge and practice relating to soil classification and land use; manuring, rotation of crops, irrigation, agricultural meteorology, tillage implements, protection, from diseases and pests, care of draught animals and grazing o cattle. The rites and ceremonies enjoined with religious practices, including the superstitions of the day, were intricately woven into the social fabric of the ancient Indians.

The *Amaraakosha* describes 12 types of land depending upon the soil, irrigation and physical characteristics. While the *Bhumivarga* describes the soils depending upon their texture, the *Vaisyavarga* describes soils fit for the different kinds of crops.

The *Brithatsamhita'* describes the seed treatment to ensure initiation of infloresence and fruit bearing of tree crops, sprouting and growth of luxurious stem and foliage.

*Agnipurana* and *Krishi Parasara* recognise the value of manuring to crops and describe the method of preparation of the compost which contribute to the fertility of the soil. Even though these authors did not investigate the precise chemical action of these manures, they certainly understood the fertilizing property and the physical effects of the manure upon the texture and its role on the water holding capacity of the soil.

The *Amarakosha* gives the names of a number of agricultural implements which were in use such as *langala, hala,* (plough), *yugakilaka* (pin of yoke), *langaladanda* (shaft of the plough), *prajana, todana* (goad), *Kotisha* (harrow), *khanitral tie* (spade or hoe), *datra, lavitra* (sickle), *yotra* (tie fastening the yoke to the plough), *medhi* (Post for threshing grain on the floor), surpa, praspotana (winnowing basket) and clani, titau (sieve). The Krishi Parashara gives a detailed description of the plough. according to this the plough is made up of the following parts: Yuga (yoke), addacalla (pins of the yoke) is a (pole of the plough), niryola (rod of the plough exclusive of the pole and the plough-share), shaula (an extra piece of wood that tightly fixed the niryola pashika (Plates), Halasthanu (strong piece of wood fixed to the niryola at the end opposite to where the ploughshare is fixed) paccani (goad made of bamboo with iron top), abandha (iron rod which prevents niryola from getting out of the pole, Yoktra (tic) and phala (ploughshare). Other implements are also mentioned like shrni (sickle), khanitra (hoe), musala (pestle), udukhala (mortar), shurpa (winnowing basket), (winnowing fan), calani (sieve) and methi (threshing post).

Manuring practices were in vogue. The Brihatsamhita says that seeds are to be sown with the addition of pork or venison into the soil and sprinkled daily with water mixed with milk. The Agnipurana recommends manuring the soil with powdered barley, sesamum and the offal matter of a goat mixed together and soaked in washings of beef for seven consecutive nights. The washings of fish should be sprinkled on these.

The farmers of the ancient days were evidently aware of the fertilizing properties of the flesh of animals, fish-washings and farm-yard manure consisting of the excreta of various animals. This was the result of practical experience and not based on scientific knowledge. Modern science has taught us that farmyard manure contains all the essential plant nutrients viz., nitrogen, phosphoric acid and potash but the ancient farmers had obtained this knowledge on the basis of empirical observations.

Land was classified into 12 types depending on the fertility of the soil, irrigation and physical characteristics in the Amarakosha. The 12 types were urvara (fertile), ushara (barren), maru (desert), aprahata (fallow), shadrvala (grassy), pankila (muddy), jalaprayamanupa (watery), kaccha (land contiguous to water), sharkara (land full of pebbles and pieces of limestone), sharkaravati (sandy), nadimatrika (land watered from a river) and devamatrika (land watered by rain).

Due recognition was given to irrigation,both natural and artificial. River water and monsoon accumulation was diverted to the fields in northern and north-eastern India whereas stored water in tanks, pools and wells as tulised in central, north-western and southern parts of India. Rules for erection of dykes and restoration of decayed dykes were also framed. During the eighth century, King Lalitaditya and during the ninth century, Minister Surya during the reign of Avantivarman in Kashmir were credited with providing canals to prevent flooding and irrigation by construction of dams and irrigation canals.

The Brihatsamhita mentions that in the Gupta period there were three harvests—summer, autumn and spring crops. The seeds sown in the summer would be reaped in Shravana (early part of autumn). The harvest was known as summer crops. The seeds sown in autumn ripended in the spring and those sown in early spring gave summer crops in Caitra or Vaishakha. The first two crops would have been rice and wheat, the last was of pulses, beans etc. Knowledgeable farmers attuned to cultivating their crops according to the season, soil, rainfall etc. Produced rice, wheat, barley,peas, lentil, pulses, spices, vegetables etc.

Kalidasa in Raghuvamsa refers to transplanted rice in Bengal. Amarakosha mentions vegetables like cucumber, onion, pumpkin and gourd; sugarcane to succeed rice and harvested during winter; cotton growing in saurashtra or Kathiawar; pepper and cardamom in the valleys of the modern Nilgiris in southern India; coconut in coastal ares of Bengali, Orissa and Madrass. In addition, reference exists to other spices like mustard, clove, ginger and turmeric; saffron, betel-nut, tamarind, sesamum, linseed: and other fibre species like silk, cotton, flax, hemp, and aloe etc., and dye crops like indigo. Amongst the fruit crops mango was the most popular.

Surapala's Vrikshayurveda written in the 10th century A.D. classifies diseases of trees into two groups viz. (1) Those arising out of internal disorders and those attacking from the outside. Internal diseases were caused by disorders of the wind (vata), phlegm (kapha) and bile (pitta) while external diseases were caused by vermin, frost etc. The methods of treatment recommended involved the administration of flesh, lymph, fat, gheer, fumigation of the tree with fumes of white mustard etc.

Breeding, feeding and maintenance of domestic animals and livestock received special attention. The Krishi Parashara outlined rules of tending cattle. Every village possessed common pasture and places of drinking water for the cattle. The agnipurana stipulates fumigation with vapurs of th e resin from Pinus deodara (*devadaru*), shorea robusta or sal resign (*guggulu*), asafoetida (*hing*) to prevent diseases. Oilcake was considered good for the nutritionof cattle.

Even though agriculture in ancient India demonstrated vigorous growth, the practices were not free from superstitions and astrological beliefs. Every even in agriculture was guided by the movement of the planets and every activity was planned to propitiate prevailing planetary influences.

Agricultural practices in India are replete with pithy sayings and proverbs, which keep guiding the agriculturists in the rural India even today.

Centuries of experience based on adoption of long trial and error methodologies in diverse topography that exists in the rural scenario of India have led India from the ancient to the medieval.

Introduction of new crops, introduction of new plants of horticultural importance brought by the foreign traders and introduction of new land tenure systems by the erstwhile rulers marked the medieval period. While the agricultural practices were more or less the same as in ancient India, the geographic distribution of the main food-crops was more or less as it is today. Renewed efforts at irrigation resulted in construction of canals of greater length in north-western part of India, whereas smaller projects were implemented in south India.

In addition to the principal food crops, cultivation of cotton was on the increase and sugarcane spread widely during the Moghul period. Bayona region was planted with indigo, Malwa and Bihar with opium, Western Ghats was initiated to coffee during the seventeenth century, Western Kashmir with saffron and fruits, the Pandyan kingdom in South with ginger and cinnamon, Malabar with cardamom, the modern Oocty with sandalwood and Kerala, south Kanara and Bengal with coconuts.

Potato and cashew from South America, pineapple from the United States of America; guagva and custard apple introduced by the Portuguese in the early 16th century; tobacco and chillies (capsicum) in the lattr half of the 16th century; tobacco and chillies (capsicum) in the latter half of th e16th century, formed some of the new crops introduced. Grafting of mango during 1550 A.D. revolutionized mango as a horticultural cash crop.

Irrigation from wells in Upper Gangetic plains; arhat or rahat (Persian wheel) in the Punjab, the Charas (bucket made of leather) in Agra and surroundings and the Denkli based on the lever principle were adopted. The tank at Porumilla constructed in 1291 A.D., and the great Anicut at Srirangam across the Kaveri by a Chola king, provided water for irrigation in the south. Tank building or restoration for irrigation purposes was treated as a benevolent act. Water lifting from wells was done with the help of a pulley wheel device fixed to a horizontal bar through which a rope is passed, to one end of which is tied the bucket, which is lowered into water and after filling is pulled by the free end by hand.

Dykes were erected and kept in constant repair to prevent flooding and prevention of entry of salt water from the estuaries during high tide and consequent waterlogging in the lowlands of Kerala an Goa to enable cultivation of rice in the valleys.

Land in the hill tops was reserved for grazing as well as for obtaining firewood. Rest of the and was assigned to joint families for raising plantations.The forests on the slopes were used for slash and burn cultivation by the farmers.

Longest canals were constructed in the early eighteenth century like the eastern Jamuna canal; Nahar-i-Bihist or Shah Nahar built by Shahjahan to bring water to Shahjahanabad in Delhi and to bring water to Lahore from the Ravi at Raipur near the Siwalik hills.

According to B.A. Keen, Director of the Imperial Agricultural Research Institute, at Pusa during 1930-'31, as supported by Dr. Anderson of Fort St. George, Madras, seed drills, with wooden bowls with three or four tubes for depositing seeds, which were introduced in England in the first quarter of the 18th Century were being used in India much earlier.

Herdsmen as expert cattle breeders were known in medieval India and the report of the Royal Commission on Agriculture (1928) attributed some of the finest breeds of cattle named after the regions where they have been bred like the Punjab, Gujarat, Mysore, Ongole etc.

**Agriculture**

The Greeks while in India has heard of the tradition that the god Dionysos first yoked oxen to the plough, and made many of the people husband men instead of nomands, and provided them with the implements of agriculture.[1] Now Dionysos has been identified with God Siva.[2] But this interpretation does not apply to all cases. We find Dionysos to be the same as Samkarsana because just as in Greece the former is associated with wine and plough so is the latter in India. Samkarsana is another name of Balarama, who is usually represented as carrying the plough on his shoulders. Therefore, reasonably we can infer that Dionysos here is not Siva but Balarama.

While classifying the people of India into seven classes or 'castes', Megasthenes observes that the husband men as a class were "far more numerous than the others."[3] From certain passages in the Jain sutras we come to know that lands and houses formed the main possession of a householder.[4] Most probably agriculture was confined to the countryside; and the husband men with their wives and children dwelt in villages and did not go into the urban settlements.[5] There was some agricultural activity in the towns or the cities also, for, the city of Aornos contained "as much good arable land as required for its cultivation the labour of a thousand men."[6] So also, the field bounds of Campa were "turned up by hundreds and thousands of ploughshares and displayed far-reaching pleasant dykes."[7] It will not be out of place here to mention that in the royal parks of Palimbothra, among the cultivated plants, there were some which received special care from the servants of the king.[8] We have instances, however, of the urban people cultivating lands which lay around their cities, that is the suburbs outside. Such lands existed round the city of Patala,[9] and also around the city to which the Malloi (the malavas) had fled for refuge when Alexander invaded their territory.[10]

Very probably in ruban as well as suburban areas agriculture was carried on, though naturally on a much smaller scale.

Agriculture was under state supervision, being superintended by commissioners.[11] It is, therefore, quite natural that the condition of agriculture, and the agriculturists was not bad.

The husband men were exempted from fighting and other public services; they could devote the whole of their time to tillage.[12] But Megasthenes perhaps did not understand the real state of things, when he says that the husbandmen were exempted from fighting; the rules of caste, in fact, did not allow the husbandmen to wield arms in battle-fields.[13]

Unlike the Assyrian kings, the Indian kings could not possibly ask the agriculturist to take up arms in times of war. For, they could not override the *dharma* of the land. Thus when Alexander asked Textiles whether he had more soldiers or husbandmen, he complained that as he was at war with Abisares and Porus, he required more soldiers than field-labourers.[14] Could he not recruit soldiers from among the field-labourers? Surely not.

Perhaps as the king did not summon the agriculturists to take up arms in battlefields, he had to maintain a large standing army at a great cost.[15] Hence in one part of the country when the soldiers were fighting out the battles of their kings, in another part the husbandmen could be seen peacefully pursuing their work without any fear of danger.[16]

The husbandmen were of a very mild and gentle disposition.[17] They were regarded as public benefactors by the people;[18] they were regarded as a class that was sacred and inviolable;[19] even an enemy, therefore, did not come upon the husbandmen at work on their land and harm them in any way; thus they were protected from all injury.[20] So we would *ipso* facto be led to suppose that in ancient India battles and wars did not affect agriculture or the people who were engaged with it. But it was not so invariably; when the aggressors came from outside India, they being foreigners,did not spare the agriculturists. Thus, for instance, when Alexander advanced upon the city of Patala the agriculturists there fled away in great terror.[21] Similar things might have happened elsewhere too.[22]

The husbandmen were dependent to some extent on neat herds, shepherds and hunters, that is, those who formed the third eats of Megasthenes, inasmuch as they freed the country from the pests with which it abounded, viz., wild beasts and birds which devoured the seeds.[23] They were also dependent on the artisans who fashioned the implements of agriculture.[24] According to Megasthenes these latter were not only exempt from paying nay taxes, but they also received maintenance from the king.[25] The king tried to improve the condition of the agriculturists by convening Great Assemblies at the beginning of every year and rewarding those scholars who discovered any means for improving the crops and the cattle.[26]

Lands for agriculture were always low-lying; they were on a level lower than that of the village dwelling site, the homestead land.[27] the agricultural lands may be classified as (i) uplands, and (ii) lowlands.[28] Upland crops and low-land crops are familiar things among the agriculturists of to-day. But one peculiar feature of the cultivation of these uplands and lowlands was that both the classes of lands seem to have been sown by the ploughmen about the same time during the rainy season.[29] If the agricultural lands were really so sown, the present practice certainly differs from the past; for, at present the uplands and the lowlands are not sown simultaneously.

There were officers, who measured the lands constantly, as was done in Egypt, evidently for purposes of the assessment of revenue.[30] This frequent measurement of lands was necessary, because many of the lauds were subject to aliuvion and diluviion.[31] These officers belonged to the Irrigation Department. They have been identified with the Rajukas of Asoka,[32] the Rajjugahakas of the Jatakas, whose duty was to measure the lands with a rajjuor rope.[33] The lands were measured by niyattanas.[34] One 'niyattana' was equal to two hundred cubits or forty thousand hastas square,that is, ten thousand yards square in English measurement.[35]

From the Jain sources were get no information regarding the communal cultivation of land. On the other hand, almost all the passages referring to agriculture seem to suggest the existence of the system of separate cultivation.[36] In this connection it is worth noticing that the Jain sutras were composed on the banks of the Ganges, that is, in Eastern India. But Nearkos tells us that among some tribes land was cultivated by families in common. When the crops were collected, each person took away a load for his subsistence throughout the year. The remainder of the produce was burnt to provide them with a reason for setting to work again and not sitting idle.[37] Nearkos evidently spoke of conditions in the Indus valley.

Agricultural labour was carried on by the house-holders themselves. The husbandmen of Megasthenes[38] were perhaps freemen householders. In the Jain literature the term gathavai includes the husbandmen.[39] The husbandmen were probably helped by their sons and dependents,[40] which led the Greek writers to think that among the people (of Moasikanos) it was a custom "to employ instead of slaves youngmen in the flower of their age, as the Creatans employ the Aphamiotai, and the Lacedemonians the helots."[41]

But sometimes the landowners had their lands cultivated by labourers,who got only a part of the produce o f the lands in lieu of wages. They were called bhagillas. These bhagillas were, however, very cruelly treated.[42]

Besides wild beasts and birds there were certain idle and cruel men who were in the habit of injuring the crops.[43] We hear that sometimes, the householder or his sons themselves set fire to the corns out of anger for some reason or other,[44] Floods were another source of danger.

The Macedonians were struck by the amazing fertility of the soil.[45] the soil near Mount Meros[46] favoured the growth of chance-sown seeds, and even laurel, spikenard, ivy and vine grew wild here.[47] The kingdom of Taxiles had a highly productive soil.[48]

If we can rely on what Diodorus says, Alexander, after overcoming king Embisaros,[49] advanced through a country of surpassing fertility.[50] This probably refers to the regions beyond the Hydaspes. Strabo says that the country beyond the Hypanis was very fertile,[51] and this is corroborated by Arrian; the inhabitants of the region were good agriculturists.[52]

About the reason of the fertilities of the soil arrian in his Anabasis says that it is due to this fact that the plains of Indian were formed from the alluvial deposits of rivers.[53] Megassthenes also holds the same view.[54] Strabo says that after the subsidence of the floods, the half-dried soil, "though scratched into furrows by any common labourer", brought whatever was planted to perfection.[55] The other parts of Indian must have been equally fertile.

The falling of the rains[56] and the melting of the snow led to the overflow of the rivers.[57] Aristoboulos says that the Indian rivers were flooded by the rains from the north.[58] It is very interesting to note that flood and inundations often made vast changes in the surface of the country. Aristoboulos says "that when he was sent on some business, he saw a tract of land deserted which contained more than a thousand cities with their villages, for the Indus, having forsaken its proper channel, turned itself into another on the left much deeper into which it burst like a cataract, so that it no longer watered the country on the right, from which it receded, for this had been raised by the inundations not only above the level of the new channel, but even that of the new inundations."[59] The excavation at Kumrahr and at besnagar have shown that sometimes whole towns were buried under neath the silts newly deposited by the floods.[60]

The floods were very heavy; and the whole country lay under water. the rivers rose to the height of forty cubits, of which twenty filled the channels to the brim, while the other twenty swept away the plains.[61] As a safeguard against the floods the towns in the Indus valley were usually bulletin 'mounds', that is, on elevated places; and in times of floods they looked like so many islands.[62] But in rural areas only the dwelling sites of the villages were on a raised level.[63] Megasthenes says that when the field's lay under water, large fishes erawled into them and at the time of the receding of waters, the husbandmen used to catch these fishes.[64] While the Macedonians encamped near the Akesines, they were obliged to shift their abodes to higher ground at the time of inundations.[65]

Floods commenced about the time of the summer solstice[66] and ceased after the setting of the Arcturus.[67]

The Greek writers do not mention that the floods caused any serious injury to agriculture. On the other hand, Strabo writes that seeds were sown even when the lands were still half-dried after the floods[68] have subsided.

Megasthenes says that famine never visited India and he adduces good many reasons for the same. First of all, there was an abundance of food-crops which included cereals, millet, pulses of different sorts, rice, bosporon and many others.[69] Next, India had a double rainfall in course of a year, on e in winter and other in summer. On account of this double rainfall there was a double crop every year; and in case one of them proved a failure, the people were always sure of the other. Besides, there was a usage, almost universally observed by the people, which might have prevented the outbreak of famines in India. In times of war the husbandmen were allowed to remain undisturbed inasmuch as the soldiers never ravaged an enemy's land with fire, nor cut down its trees.[70] Megathenes probably meant a general and protracted famine; for, famines were not quite unknown in ancient India. In Antagada-dasao the 'famine-food' for the Jain monks is mentioned.[71] But then towards the end of Gandragupta's reign, a terrible famine broke out in Magadha, which extended over twelve years. Such a prolonged famine must have resulted in serious economic dislocation throughout the country. It is very strange that Megathenes does not make mention of this great incident. Possibly he left India before the famine concurred, that is, before about 302 B.C.

**Rainfall**

Just as in modern times, so also in the past, India enjoyed the benefits of a double rainfall. But Nearkos wrote that the plains of Indian were watered with rain in summer, but were without it in winter;[72] obviously, Nearkos wrote it from his experience. During the winter the Macedonians stayed in the scraggy heights of the Hindukush, when mountains were covered with snow, and rains were not seen.[73] During the course of their ten months' sojourn, which occupied the whole of the late autumn, the winter and the following spring and the summer they saw no rain.[74]

As to what aristoboulos said about the rainfall of India there has been much misunderstanding. The followers of Aristoboulos in later times observed that the plains of India were not watered by rain.[75] In a passage where Strabo quotes Aristoboulos, it is said that the mountains and the regious at their foot received rains and snow, while the plains had neither.[76] By "regions at the foot of the mountains" Aristoboulos probably meant Taxila and the surrounding country, and by 'plains' he meant that part of India which lay between the Hydaspes and Patalene. For, he refers to the absence of rains in this part of India.[77] But he says that in the country beyond, rains and snow fell and lands were easily cultivated.[78] Here he is probably referring to the Upper Punjab and the mountainous regions of northwestern India or to the country west of the Indus between the Hydaspes and Patalene.[79]

In the mountainous regions to the north-west of India rains were very heavy. In the country round about Nysa, modern Naghenhar, rains used to pour down so heavily that the bunches of grapes fell off in the vineyards.[80] Aristoboulos writes that the Indian rivers were flooded by the rains from the north.[81] In the mountainous countries of Paropanesene, Aspasioi and Assakenoi[82] rains set in early in spring and poured down day and night in torrens without intermission.[83] When the Macedonians were in Taxila, rains fell for the first time.[84] But the truth of this statement is doubtful; for they had already experienced heavy rainfalls in Paropamisadai, Aspasioid and Assakenoi.[85] During the course of their march eastwards from Taxila, the Macedonians so that the whole country between Taxila and the Hypanis was being watered by heavy rains.[86]

Kathiawad and the neighbourring regions were also subject to heavy rains;[87] we hear that the lake Sudarsana, which was built during the reign of Candragupta Maurya, thrice burst out on account of heavy rains.[88] The *Arthasastra* contains certain details about the rainfall in the different parts of India,[89] e.g., Anupa received twenty four *dronas* of rainfall, Avanti 23 *dronas,* Jamgala (probably Kurujamgala) 16 *dronas,* Asmaka[90] 13½ *dronas* and the western countries and the borders of the Himalayas got an immense quantity of rainfall.[91]

The following passage from the *Acaranga sutra* gives a very brief description of the country during the rainy season; "when the rainy season has come and it is raining, many living beings are originated and many seeds just spiring up, the roads between (different places) contain many living beings, seeds etc.... the foot-paths are not used, the roads are not recognizable."[92] From this it appears that the eastern Gangetic plain received a fair quantity of rain.

India, as Eratosthenes states, received her summer rains from the Etesian winds and from the vapours which arose from the rivers.[93] Rains used to set in early in spring, and lasted till the rising of the arcturus, that is, the beginning of autumn.[94] So the rainy season extended over about six months. But this was the case only in the Indus valley, that is, western India.[93] In the Gangetic plains rains lasted for four months, and ceased only in the beginning of Margasirsa.[96] In other words, in the eastern part of India rains set in the beginning of Sravana and Lasted till the end of Kartika or the beginning of Margasirsa. From the Girnar inscription of Skandagupta we come to know that the rainy season followed the summer and that the country witnessed heavy rains in the month of *Prosthapada,* so that the period of rains roughly coincided with that of Gangetic plains.[96]

Lands were irrigated with the waters of many rivers.[98] Quintus speaks of the river Ethimanthus which was used for irrigation by the people on its banks.[99] We are not, however, certain whether the waters of the Indus and tributaries were used for purposes of irrigation.[100] Some rivers supplied water even for the nurture of garden vegetables.[101] In fact, there was an extensive system of irrigation, though, of course there might have been lands which were cultivated without the help of irrigation.[102] In 1914 the remains of an irrigation canal of the Maurya or possibly pre-Maurya period were exhumed by Dr. Bhandarkar at Besnagar. The canal was seven feet broad and five feet and six inches deep.[104] King Mahapadma Nanda extended several channels *(panadi)* for purposes of irrigation from Tanasuliya vada to the capital town of Kalinga. The Archaeological Department has not been able to find out any trace either of the reservoir or of the channels.[104] This, however, shows that extent of care and attention paid by the ancient kings to matter of agriculture and irrigation.

Under the Mauryas also irrigation was under state supervision.[105] There was a special Irrigation Department maintained by the king Megasthenes wrote, "some superintend the rivers, measure the lands, as is done in Egypt, and inspect the sluices, by which water is let out from the main canals into their branches, s that every one may have equal supply of it,[106] From this Vincent Smith infers that a water-rate must have been levied, and that there was 'a regular system of canals.'[107]

The Girnar rock inscription of Rudradaman tells us that Pusyagupta the Vaisya, who was Candragupta's governor of the western provinces (rastrika), built a reservoir by making an embankment across a gorge of the hill for purposes of irrigation and named it 'Sudarsana.' But for some reason or other he could not construct the irrigation channels, and these were done by Yavanaraja Tusaspha, the viceroy of Asoka.[108] The scheme was an ambitious one and it required much time and labour. Begum during the reign of CAndragupta it was completed only in the reign of his grandson Asoka.

In consequence of a double rainfall and an extensive system of irrigation the soil bore two crops in the course of a year.[109] In the winter season wheat,[110] barley, pulses and many other esculents were sown;[111] and in the summer or rainy season rice, bosmoron, sesamum, millet and flax.[112] Strabo ascribes as double crop also to the fruits.

Ploughing was perhaps carried on with the help of oxen. In the country of the Aspasioi Alexander the Great captured about 230,000 oxen. Out of them he selected those better in size and beauty and sent them to Macedonia to be employed in agriculture.[113]

Probably for a deeper ploughing some people employed elephants in agriculture; they regarded them as forming the main part of their stock in cattle.[114]

From what has been said above it can easily be inferred that big ploughs were employed in agriculture. In the *Uvasagadasao* the teeth of a pisaya have been compared with ploughshares.[115]

During the growth of the plants the cultivators had to take great care of the fields, as the weeds that grew by the side of the plants had to be removed.[116] IN the Jain literature[117] where the cutting of the crops when ripe is referred to there is no mention of the sickle. Sickels mush have been used, for at taxila and Sarnath such implements were found in course of excavations. Affluent landowners had a large number of carts for carrying the grains from the fields to the born-door;[118] they also possessed a large number of boats which were perhaps employed for the same purpose.[119] It may be added here by way of a explanation that probably the rich landowners had large acres of land, which lay a long way off their houses or their barns. Then the seeds were dried in the sun.[120] Threshing and husking processes have also been referred to by Strabo. He says that rice was husked in the same way as barley.[121] In the *Acaranga Sutra* the winnowing of grains is mentioned.[122] The winnowing sieve was called suppa.[123] besides winnowing, we hear of the grinding of grains on a rock (silac) or 'on a piece of clay.[124] The granary was called *pinnagapindi*[125] or *khala*.[126] Now from the Vedic literature[127] we come to know that in the granary bundles of grains were beaten or trampled upon.[128] There were storehouses (*kosthi or kolejja or kosthagara*[129]) for the preservation of grains.[130] The kings also used to keep storehouses for emergencies.[131]

*A.K. Bose*

## REFERENCES


1. Arr. Ind. VII.
2. M'Crindle, VI. p. 64 fn.
3. Meg. Frag. 1.— Doid. II. 40. ef. Meg. Frag. XXXII.— Strabo XV. I.40. "The second caste consists of the husbandmen, who form the bulk of the population"; ef. Meg. frag. XXXII.; Arr. Ind. XI. "The second caste consists of the tillers of the soil, who form the most numerous class of the population". Megasthenes evidently spoke of India so far as it was under Candragupta Maurya. Cf. Meg. Frag. LVI; Pliny. H.N.VI. ".... Some till the soil...." ef. Meg. Frag. LVI. B.; Solin. 52.6.17; ef. Arr. Ind. XVI.
4. Uttar. XIII. 24; Ibid., IX. 49; ef Ac. Su., 1. 2. 3. 3; Ova., I; Ac. Su., II. 7. 2. 4-6.
5. Meg. Frag. I; Diod.II. 40; ef. apast. 1. 32. 21; also Baudh. II. 3.6.33. Vide also Camb. IIist. Ind., vol. I, p. 237; also the section on towns.
6. Arr. Anab.IV.XXVIII. Vide also Itin. Alex. 108 & 112.
7. Uv. sec. 1. (In the Jain sacred works the descriptions containmuch that is exaggerated).
8. Aclian. XIII. c. xviii.
9. Arr. Anab. VI. xvii.
10. *Ibid.*, VI. vi.
11. *Ibid.*, VI. xii. cf. Meg. Frag. XXXIV; Strabo XV. 1. 50-52.
12. Meg. Frag. I; Diod. II. 10.
13. *Ibid.*, I; Diod. II. 41.
14. Hist. Alex. Q. Curt. Ruf. VIII. xii.
15. Meg. Frag.I; Diod. II. 41; *Ibid.*, XXXII; Arr. Ind. XII; *Ibid.*, XXXIII; Strabo XV. i 47.
16. *Ibid.*, XXXIII; Strabo XV. i. 40; Ibid., I; Diod. II. 36; Ibid., XXXII; Arr. Ind. XI.
17. *Ibid.*, XXXIII; Strabo XV. i. 40.
18. *Ibid.*, I; Diod. II. 40.
19. *Ibid.*, Diod. II. 36.
20. *Ibid.*, Diod. II. 40. But some wicked persons intentionally injured the crops inthe field.
21. *Ibid.*, XXXII; Arr. Ind. XI; Ibid., XXXIII: Strabo XV. i. 40.

22. Many, for example, were sold into slavery, ef. Hist. Alex. Q. Curt. Ruf. IX. iv.

23. Meg. Frag. I; Diod. II, 41, Ibid., XXXIII; Strabo XV. i. 40.

24. *Ibid.*, I; Dio. II. 40; Ibid., XXXIII; Strabo XV. i. 40.

25. *Ibid.*, I; Diod. 40.

26. Strabo, XV. i. 39.

27. *Ibid.*, XV. i. 45.

28. *Ibid.*, XII, 12.

29. Ibid., This is quite evident from the passge itself. Here no reference is made to any subsequent cultivationof either the upland or the lowland crops.

30. Meg. Frag. XXXIV; Strabo XV. i. 50-52.

31. Vide Ghosal, *Hindu Revenue System*, p. 172.

32. P.E. Iv. vide also Raichowdhury, Pol. Hist. Anc. Ind. (2nd edn.), p. 199. King Hastipala had an assembly-house for his arajjukas.

33. The rajus were probably something like the modern chains.

34. Uvas. I. 19.

35. See Hoernle's notes on Uras. I. 19.

36. Ac. Su., II. 10. 10.; Uvas. I. 19.

    I.H.Q., June, 1934.

37. Strabo, XV. i. 66. The crops were burnt for purposes fo serving as manure for the next sowing.

38. Meg. Frag. I; Diod. II. 40; *Ibid.*, XXXII; arr. Ind. XI; *Ibid.*, XXXII; Strabo XV. i. 40.

39. Uvas., 1.2. There are innumerable references to the terms in Ac. Su., Uttar. Kalpasuttra and many other sacred books of the Jains.

40. Cf Ac. Su. I. 2. 5. I, wher ethe joint family members have been mentioned. Vide also Ac. Su., II. 10. 10.

41. Strabo XV. i. 34.

42. Bhagilla (bhagila), possibly an equivalent of bhagika.See Sutrakrt. II. 2, 63. We are not however, certain whether bhaga means a sixth part of the produce, as Prof. Jacobi thinks.

44. The crops named are kalama, masura, sesamum, mudgam beans, nispava, kulattha, alisanda, clamiccha, — Sutrakrt. II. 2.63.

44. Sutrakrt. II. 2.44.

45. Aelian, HIst. Anim. XII XXXII; Meg. Frag. I; Dicd. II. 35; Pliny VII. 2.

46. Mount Meros is modern Mar Koh. It lay near the city of Nysa (modern nanghenhar, if Nysa be Nagra of Ptolemy), Nysa is ancient Noacha of Taxila —Silver Scroll Inscription.

47. Hit. Alex. Q. Curt. Ruf. VIII. X.
48. Plutarch's Life of Alex., ch. IAX. See also Marshall's Guide to Tarila, p. 2.
49. Possibly kng of Abhisara is meant.
50. Bib. Hist. Diod. Sic. XVII. xc.
51. Strabo XV. i. 37.
52. Arr. Anab. V. vi.
53. *Ibid.*
54. Meg. Frag. II.—Arr. Exped. Alex. V. 6. 2-11.
55. Strabo XV. i. 13.
56. This clearly refers to the rains and the melting of the snows in summers; Strabo. XV. i. 13.
57. Strabo, XV. i. 17; Arr. Anab. V. ix.
58. Strabo, XV. i. 19.
59. *Ibid.*
60. A.S.R., 1913, 14; 1912, 13.
61. Strabo XV. i. 18.
62. *Ibid.*
63. Ibid., XV. i. 45.
64. Meg. Frag. LIXZ; Hist. Anim. XVI. 19.
65. (In summer solstice). Strabo XV. i. 19.
66. Rains fell inabout the summer solstice, i.e. about the setting of the pleiades.
67. Strabo XV. i. 18.
68. *Ibid.*
69. Meg. Frag. I; Diod. II. 36; cf. Strabo XV. i. 13; IBid. XV. i. 20; Arr. Ind. V.
70. Meg. Frag. I; Diod. II. 36. For a recent explanation of this assertion of Megasthenes see Radhanraman Gangopadhyaya's Some Materials for the Study of Agriculture and Agriculturists in Ancient Indian,' pp. 112-126.
71. Meg. Frag. I; Diod. II. 36; Arr. Anab. V. IX.
72. Strabo, XV. i. 18.
73. *Ibid.*, XV i. 17.
74. *Ibid.*
75. *Ibid.*, XV. i. 24.

76. *Ibid.,* XV. i. 13; cf. Ktesias. Frag. I; Bib. LXXII. p. 144.

77. Strabo, XV. i. 19. and 26.

78. *Ibid.*

79. For to the east there is the desert of Rajputana which was well known to the Maccdonins (Herod. III. 98). Here inteh heat of the sands, there could nto have been any snowfall, nor anyrains. But in the country west of the Indus we have the mountains of Beluchistan where there were possibly falls of rains and snow. Cf. Vincent smith, E. H. I., 4th. edn., pp. 40-41.

80. Meg. Frag. XLVI; Strabo XV. i. 6-8.

81. Strabo XV. i. 19.

82. Paropanesene is the Hindukush. The territories of the Aspasioi and the Assakenio are to be located inteh valleys fo Bajaur and the Panikora.

83. Strabo, XV. i. 17.

84. *Ibid.*

85. *Ibid.*

86. *Ibid.*

87. EI., VIII., 36; Fleet's Gupta Inscriptions, pp. 56-65.

88. *Ibid.*

89. In the Arthasastra we find that the raingauge (varsamana) was used for measuring the rains in those days. It was a very simple contrivance. In front of store-hourses, there were bowls (kunda) the mouths of which were wide as an aratin, that is, twenty-four angutas; and these served as raingauges.

90. Either the Aspasioi and the Assakenoi are meant or the Asmaka kingdom of Deccan.

91. Artha. S., II, 24.

92. Ac. Su., II. 3.1.1.

93. Strabo XV. i. 13.

94. *Ibid.,* XV. i. 17.

95. For, the knowledge of the Greeks about India was confined to the western part of our country.

96. Ac. Su., II. 3. 1. 4.

97. Fleet's Gupta Inscriptions, pp. 56-65.

98. Meg. Frag. I; Diod. II. 36-37. This explains why someof th eofficers of the Irrigation Dept. had to superintend the rivers. cf. Dion Chrys. Orate. XXXV. 434.

99. Hist. Alex. Q. Curt. Ruf. VII. ix.

100. Meg. Frag.I; Diod. II. 37. The most notable of th etributaries of the Indus were "the Hupanes, the Hudaspes and the Akesines"; Cf. Arr. Ind. IV> "We ought not ...... canals".

101. *Ibid.*

102. Meg. Frag. I; Dio. II. 35. The expression "greater part of the soil is under irrigation" implies that there were probablylands, which did not require to be irrigated.

193. A.S.R., 1914-1915, II, pp. 69-70.

104. J.B.O.R.S., 1917, pp. 425-472.

105. Meg. Frag. XXXIV; Strabo XV. i. 50-52.

106. *Ibid.*

107. Smith's E. H.I., 4th edn., pp. 138-140; Oxford History of Indian, 2nd edn., p. 91; Asoka, (Rulers of India Series), p. 95.

108. El., VIII, 36.

109. Meg. Frag. I; Diod. II. 35-36; Strabo XV. i. 13 and 20. *Ibid.*, XVI. iv. 2.

110 Meg. Frag. I; Diod. II. 36; Strabo, XV. i. 13.

111. Strabo, XV. i. 20.

112. *Ibid.*, XV. i. 13; Meg. Frag. 1; Diod. II. 36.

113. Arr. Anab. IV, xxv; cf. Itin. Alex. 105.

114. Meg. Frag. LVI; Pliny N. H. VI. See also N.G. Banerji's Economic Life and Progress in Ancient India, ppp. 116, 121. Prof. Banerji's suggestion is that te employment of oxen even to the number of 12 implies either a deep ploughing or the hardness of the soil. But the theory of the existence of a deep ploughing is more probable than the theory of the hardness of the soil. The alluvial soil of North India is naturally soft, and some 20 years back, Dr. Spooner proved that the soil of Patna district has been extremely soft from time immemorial (ASR., 1912-13. II).

115. Uvas. II, 94.

116. Sutrakrt., II. 2.11.

117. Anuttar.

118. This is based on the explanation of the term Samvahanichim as given in the commentary. Though the commentary is of a later date, yet it seems that the explanation given there is probable, and has been taken from earlier commentators. For parts of the commentaries on Jain works are traditional, being handed down from teacher to papil.

119. Uvas. I. 121; vide also no. 119.

120. Anuttar.

121. Strabo XV. i. 18.

122. Ac. Su. II. 1. 6.8.

123. Uvus. II. 94.

124. Ac. Su. II. 1. 6.8.

125. Sutrakrt., II. 6. 26.

126. *Ibid.*, II. 2. 44.46.

127. N.C. Banerji's Economic Life and Progress in Ancient Indian, p. 115.

128. *Ibid.*

129. Uttar. XI. 26

130. Ac. Su., II, 1.7.2.

131. Sohgura Inser. in IA., 1896.

# 4

# Cultivation in Ancient India

The earliest mention of agriculture is found in the *Rgvedic* hymns. By the time the hymns were composed, the ancient Aryans settled down in 'Sapta-Sindhu'[1] and the ancient Punjab, and the *Rsis* invented agricultural implements and inaugurated sacrifices in honour of Indra for timely rains for carrying on successful agriculture operations. Cattle, besides being sources of nutritious food such as milk, curd, clarified butter, etc. and sometimes meat also, were yoked to the plough for cultivation and their dung was employed as manure. Professors Macdonell and Keith say that *sakrt* in the *Rgveda* denotes dung and "it is clear that the value of manure was early appreciated.2 the *Vedic* Aryans therefore highly valued their cattle and their solicitude for their safe-keeping is clearly indicated by the following passage;[3]

"Construct the cow-stall, for that is the drinking place of your leaders (the gods); fabricate armour, manifold and simple; make cities of iron and impregnable; let not the ladle leak, make it strong."

*Yava* and *dhanya* (barley and paddy) were the staple crops from the very beginning, *Yava* being a spring crop might not have required much rain, but for *dhanya* a regular rain-fall was needed for which Indra was invoked. Besides this natural supply of water, the Vedic Aryans used to have recourse to irrigation by artificial wells and broad channels along which water was directed to different parts of the *ksetra*.

The following Rgvedic Passes[4] about agriculture are significant: "Harness the ploughs, fit the yokes; now that the womb is ready, sow the seeds therein and, through our praise, may there be abundant food; may (the grain) fall ripen towards the sickle".

"Set up the cattle troughs, bind the straps to it; let us pour out the water of the well, which is full of water, fit to be poured and not easily exhausted."

## The Atharvaveda

In the Atharvaveda, we find that the value of natural manure of animals in the field was perfectly understood,[5] the Atharvadeda VI, 50 mentions the animal enemies of corn such as the locusts, the rat, the bars, etc. and invokes the diving power for their destruction.

## Other Treatises.

By the time the Kautiltya Arthasastra was composed, agriculture became an important department of the government under the charge of a special officer well-versed on the subject, and attained a stage of perfection astonishing for that age. The Arthasastra, though essentially a treatise on royal polity and art of government, deals with agriculture under an important sub-head of *Krsi-tantra.* The Brhat-samhita, the Gargya-samhita and the Agnipurana also incidentally deal with plant physiology, selection of soil, manuring, cultivating under favourable meteorological conditions, collection and treatment of seeds, sowing, planting, cutting, grafting, nursing and reaping. Much of the art of cultivation as it existed in ancient days can also be gleaned from a hand-book written by Parasara bearing the title of *Krsi-samgraha,* This small treatise supplemented with Khana's maxims throws a flood of light on the perfection attained in the art of agriculture in ancient India.

The account is given under the following:

1. Soil and its classification.
2. Meteorological observations leading to forecasting of rain-fall.
3. Keeping of cattle needed for cultivation.
4. Preparation and application of manure.
5. Collection and treatment of seeds.
6. The construction of agricultural implements.
7. Ploughting,
8. Sowing and planting, and treatment of plants etc.
9. Reaping and storing.
10. General instruction to cultivators.

The body of ancient Indian literature is vast, being a store-house of human experiences gathered in the curse of ages. I cannot certainly claim to have ransacked it all, nor is it possible to do full justice to a subject so wide and so important in an article like this, but I do hope that it would help to draw the attention of new investigators to this line of research.

## I

## SOIL AND ITS CLASSIFICATION

### Classification

According to fertility, soil is mainly divided into two classes: *Urvara* (fertile) or Usara (sterile). *Urvara mrttika* is again sub-divided into different kinds according to their peculiar fitness for the cultivation of different kinds of crops, for instance, *Yavya tila, vraiheya saleya, maudgina,* (desert). The soil watered by river and that watered by rain are respectively called *nadi-matrka and deva-matrka.*

**Selection**

The exact chemical composition of different kinds of soil might not have been known to the ancient authors but from the following typical maxims[6] of Khana, it becomes clear that they made extensive experimental observations about them and obtained a masterly knowledge regarding their characteristic suitability for the cultivation of different kinds of crops.

"Sandy soil is suitable for the cultivation of Aus paddy and clay soil for that of jute.

"Your expectation will be fulfilled if you cultivate *patrol* (Trichosanthes dioica) in sandy alluvial soil."

"Potatoes thrives well, if cultivated by the side of a bamboo-grove."

If arum is sown on the bank of a river, it grows three cubits long."

In the Arthasastra, we find an enumeration of the suitability of different lands for the cultivation of different crops.

"Lands that are beaten by foam, i.e., banks of rivers etc. are suitable for growing Valliphala (pumpkin, gourd and the like); lands that are frequently overflown with water (parivahanta) for long pepper, grapes (mrdvika) and sugar cane; the vicinity of wells for vegetables and roots; low grounds (haraniparyanata—moist bed of lakes, *com.*) for green crops; and marginal furrows between any two rows of crops are suitable for the plantation of fragrant plants, medicinal herbs, khuskhus roots (ursira), hira (r), beraka (?) and pindalaka (lac) and the like.[7]

**Indispensability of water**

That water is essential for germination, that the soil must be sufficiently moist and the moisture maintained until the plant is established is fully recognized in the following verse from the Krsi-samgrahah.[8]

"Rain is essential for cultivation and the latter is essential for life, so one should first acquire carefully the knowledge about rain-fall."

## II

## METEOROLOGICAL OBSERVATIONS LEADING TO FORECASTING OF RAIN-FALL

Parasara in his Krsi-samgrahah deals elaborately with meteorological observations leading to predictions about rain-fall or scarcity of water and consequent abundance or dearth of crops. The following quotations will bear proof.

**Influence of the planets etc.**

"If the Sun be the ruler of the year (warshdwipa), the rain-fall is moderate. If the year comes under the influence of the Mars, the rain-fall is scanty, and it is abundant if the year comes under the influence of the Moon, the Mercury, the Jupiter and the Venus. If the Saturn rules over the year, there will be drought and the earth will be full of dust." 13, 14.

"The earth is partially drenched under the influence of *Avarta,*[9] and wholly drenched under the influence of *Samvarta.* Under *Puskara,* the rain-fall is scanty and under *Drona,* it rains in abundance. 24-26.

"If the Moon enters Gemini, Aris, Taurus or Pisces or the Sun enters Cancer, there will be 100 *adhaks*[10] of rain; there will be 50 *adhakas* of rain when the Sanenters Leo or Sagittarius, 80 *adhaks* when it enters Virgo or Capricorn, and 96 *adhaks* when it enters Cancer, Aquarius, Scorpion and Libra. 28, 29.

"Of this 50% will fall on oceans, 30% on mountains and 20% on land."

"Rain-fall is sure during the transit of the Mars and the Saturn from one zodiac to another and the Earth is fed with heavy showers just before the transit of the Jupiter." 69.

"If the Jupiter enters the asterism known as *Citra,* clouds burst out in the same way as water bursts out from a broken pitcher, but if the Jupiter enters *Svati,* it des not rain at all in spite of thickest clouds."

**Monthly indications.**

"Heavy showers will fall on the earth if the sky is cloudy and lightening flashes in the western sky during the bright half of the moon (sishe pache) in the month of *paus,"* 35.

"If any day in the month of *Paus* becomes rainy or foggy, seven months hence, it must rain cats and dogs in the corresponding day." 36.

"The year in which clous are seen or rainfalls on the seventh day of the full-moon in the month of *Magh,* is a blessed one and yields plenty of crops."

"If on the seventh day of the new-moon in the month of Magh, it rains under the influence of *Svati* or strong inds blow or rain-laden clouds roar in the day-time or the sky becomes covered with lightening flashes and the moon and the stars become envered with lightening flashes and the moon and the stars become enveloped, then there will be an excellent rain-fall up to the month of *Kartika.*" 38.

"If the seventh day of the full-moon in the month of *Magh or Falgun* or the 3rd day of the full-moon in the month of *Chaitra* or the first day of *Vaisakh* be a day of thunder-storm and rain, then in that year, there will be no scarcity of rain and the earth will yield crops in abundance." 39, 40.

"If it rains any day in the month of *Agrahayan or Magh or Falgun,* then, in the seventh month from that day, the earth will receive a heavy shower," 41.

"If the first day of the full-moon in the month of *Chaitra* falls on a Sunday, then there will be a moderate rain-fall; and if it falls on Monday, the earth will be flooded, But if it is a Tuesday, it will be year of drought and dearth. If it is Wednesday, Thursday or Friday, the earth will be full of corns and if Saturday, even the sea gets dried up and the earth becomes dry and dusty." 42, 43.

"There will be a moderate rain-all during the year, if it rains under the auspices of *Citra* in the first half of the month of *Chaitra,* and a heavy rain-fall if it rains in the middle or the latter part of the month." 44

"If under the auspices of the asterisms *Svati, Citra and Visakha,* the sky be cloudless in the month of *Yaistha* or it rains in the month of *Sravan,* then there will be a pretty good rain-fall that year." 55.

"Rain-fall will be good throughout the year, if it rains on the (the day of the full-moon of *Asadh.* If otherwise, thee is no prospect of rain that year." 57

"In the year in which it does not rain under the auspices of *Rohini* in the month of *Sravan,* people will be in distress and cultivation a failure." 60, 61.

In the following verses Parasara gives *another method* predicting the rain-fall for the year.

He directs to take a rod and write an mantra on it two hundred times on the first night of the full-moon in the month of *Vaisakha* and then erect it on the bed of a flowing river and mark where the water level stands. It the water level is found not to have changed the following morning, then the rain-fall will be the same as in the previous year. The rain-fall will be more or less than that of the previous year according as the water-level goes above the mark of below it. 45-49.

That the ancient people could forecast weather conditions from the *direction of the wind* may be guessed from the following verses.

**Prediction from wind directions.**

Parasara[13] says, "the rain-fall will be good if in the month of Paus wind blows from the north or from the west, and the rain-fall will be scanty if the wind blows from the east or the south. It does not rain at all, in the case of the wind not blowing unidirectionally and strongly." 32

"It rains well if on the full-moon day of *Asadh,* wind blows towards the east. It means damage to crops if it blows towards the south east, scanty rain-fall if it blows towards the south and destruction of crops if it blows towards the south-west. A westerly wind means rain-fall and a north-westerly, storms. If the wind is northerly or north-easterly, it dorebodes abundant harvest." 56

**Prediction of immediate rainfall or scarcity.**

As for the prediction of immediate rain-fall, Parasara[14] gives the following.

"There will be an immediate rain-fall, it water-spouts are seen at or near water or if ants suddenly line upwards after collecting their food or if frogs crock suddenly." 63, 64.

"If cats, weasels, snakes and other creatures that live in holes and pits and *Sarabhas* (a kind of deer) run excited or if boys build clay bridge-ways in the middle of a path or if peacocks go on dancing, then rain must be falling immediately." 65, 66.

"Clouds will be presently pouring down, if sudden pain arises in the body of a sprained or rheumatic man or if snakes climb to the top of trees or if water-birds dry their wings in the sun or if buzzing sounds are heard in the sky." 67, 68.

Signs of drought are given in the following passages of the Krisi-samgraha.[15]

"There will be drought if the Mars goes into *Uttarafalguni, Uttarasadha, Uttarabhadrapada, Sravana, Hasta, Mula, Iyestha, Krttika, or Magha.* Even the sea is dried up if the Moon closely follows the Mars, and drought is impending if the venus enters *Citra.* The earth becomes scorched up (for want of rains) if the Mars goes into Leo and the ocean becomes dry if it is in conjunction with the Moon." 73-75.

In the Arthasastra also we find that "A forecast of such rainfall can be made by observing the position, motion and pregnancy (*garbhidhana*) of Jupiter (Brhaspati), the rise, setting and motion of Venus and the natural and natural aspect of the Sun.16

Some of the aphorisms of Khana as to rain-fall and scarcity of rain and their influence on harvest are given below:

**Khana's Maxims.**

"Rainy day and cloudy night predict a drought and warm *Paus* and cold *Vaisakh* predict a heavy downpour in *Asadh'.*

"It rains cats and dogs, if stars are seen within the moon's halo. The bigger the corona of the moon, the nearer is the rain-fall."

"Rainbow in the western sky (i.e. in the morning) predicts the scarcity of rain, while that in the eastern sky (i.e., in the afternoon) means a heavy shower."

"If it rains in the month of *Agrahayan* even the king is driven to begging; and if it rains in *Paus*, even the husks are sold high."

"If it raise towards the end of *Magh*, glory be to the king and holy is the land,and if in *Phalgun, cina and kaon* (grains) grow in plenty."

"Drought in *Jyaistha* and showers in *Asadh* lead to crops too abundant for the earth to bear."

"If it rains in *Asadh*, paddy grows in plenty."

"There will be an abundant harvest if it rains in the month of *Chaitra*."

"Aus paddy doubly grows, if it rains in the beginning of Vaisakh."

"Khana says to the cultivator that if fleecy clouds be followed by wind in the full-moon day of the month of *Kartik*, winter crops grow plentifully; but if there be cloud and rain in the night, it is fruitless to go to the field (i.e., to expect any harvest).

"The year in which it rains in the (the day of the full-moon in the month of *Asadh*, the heron will walk over the very bed of the sea, If it drizzles on that day, it will be followed by heavy showers throughout the year to the extent of making fish inhabit the top of mountains (i.e., the earth will be flooded). If it rains now and then, there will be a rich harvest and if the sun sets under a clear cloudless horizon, the crops will not grow at all."

## III

## KEEPING OF CATTLE NEEDED FOR CULTIVATION

From the earliest times, cattle were highly valued and cared for. Directions for their safe keeping, etc. are given in the following passages:[17]

### Erection of Cow-stall and Sanitation

"A cowshed fifty five cubits square is conducive to the health of cattle: but it is injurirous for them if it is erected in the direction of Leo." 86.

"If one through ignorance or carelessness erects a cow-stall when the sun is in Leo, i.e., in the month of *Bhadra* or keeps rice-washings, hot starch, fish-washings, cotton or husks in the shed, then all his cattle shall perish." 87, 88.

"The same thing happens if goats are tied or broom-stick, husker, or spoiled food kept, in the cattle shed."

"If the cattle-shed be strong, clean and free from dung, the animals have a healthy growth in spite of poor food; but if they have to come out of the shed besmeared with their excreta, even highly nutritious food cannot help their growth."

84, 85.

"Fortune abides in the cow-shed that is free from filth, dust and human excreta; misery visits the shed and the inmates weep if it is not lighted every evening."
93, 94.

**Employment of Cattle**

"The animals are to be so employed in cultivation, that it may not hurt them, or the corns will be unsuitable for any holy purpose, and even if they grow in plenty, they cannot bring prosperity to the cultivator." 81, 82.

"Urine of the domestic cattle should never be used for cleansing filth, for that is (indirectly) harmful to them." 90.

"One who is solicitous about their well-being should never even erroneously give their dung to others on Saturdays, Sundays and Tuesdays. These three days (of the week) are to be avoided—particularly Saturday and Tuesday; otherwise, the cattle will meet with harm." 91, 92.

"One who yokes 8 cows to a ploughis a poius man and one who yokes 6 is a business man. Cruel are those who employ 4, and those who employ only 2 are beef-eaters." 95.

Parasara directs that an *annual ceremony* in honour of cattle be performed in the following way:

"On the first day of the full-moon in the month of *Kartik,* a cow is to be worshipped after typing round its neck a green creeper and rubbing its body withhold turmeric paste. Then a beautiful bull is to be anointed with sandal and *Kunkum* (saffron) and the cultivators after having dressed themselves with ornaments and goads in their hands should lead the bull round the village with the accompaniment of dances and music." 91, 99

"On the first day of the month of *Kartik,* cattle should be rubbed, with a mixture of oil, turmeric and *Kunkum* powders, their bodies being marked with a piece of hot iron and the hair at their tails and ears clipped. That is sure to make them free from diseases and keep them in health and cheer throughout the year."[18]
100,101,102.

The above shows how anxious the ancient people were for safe and healthy keeping of their cattle and what a sanctity they attached to the cow. The custom of marking cattle is as old as the time of the Rgveda when the ears of cattle were marked to indicate ownership, and the special sanctity that the cow acquired is early "shown by the name *agony* (not to be killed) applied to it in several passages[19] of the ancient scripture.

## IV

## PREPARATION AND APPLICATION OF MANURE

"In the month of *Magha,* a dung-heap is raised by a spade, dried in the sun and made into smaller balls. In the month of *Phalguna,* these are placed into holes dug for the purpose, and afterwards scattered on the field at the time of sowing. The paddy-plant only grows without manure, it does not bear fruit." 107, 108, 109.

The ancient Indians did not apparently know the use of chemicals as artificial fertilizers; they have come into use only about two hundred years ago. Besides bones, flesh of animals, fish-washings, vegetable and animal products etc., the manure that they primarily used consisted of the excreta of various animals mixed with litter which absorbed the urine and kept the animals clean. Whether they knew the exact chemical composition of the yard-manure is not known, but they certainly appreciated its fertilizing property and also its physical effects upon the texture and water-holding power of the soil. It is only a modern discovery that the farm-yard-manure contains all that is necessary for the nutrition of plants, viz., nitrogen, phosphoric acid and potash. Nitrogen compounds are the chief fertilizing elements in the manure, but the nitrogen is inevitably lost to a certain extent. The loss can be minimized only if the dung-heap is not disturbed; for any disturbance causes rapid fermentation of the liquid portion of the manure, viz., urine, with a consequent increase in the evaporation of ammonia. The direction of Parasara to keep the dung-heap undisturbed up to the month of *Magha,* i.e., for ten months of the year is thus significant. Again, the sentence, — "Roudre samsocia tan sarve krutva guruvokarupe ram" is also significant, for if any easily fermentable material still remained in the active form, it should be got rid of by drying. This process also reduces ammonia which would otherwise be injurious to the seeds and the tender roots of plants. The direction "gatra putva midyardhit" is also very important inasmuch as the manure, as it decomposes under the earth, increases the stock of humus which oxidises and tends to decrease in the open air, so that when the manure is scattered over the field, it is comparatively rich in humus that contributes greatly to the fertility of the soil. Knowledge of manuring at the time was probably a result of extensive practical and not scientific observations.

There are also many other writers who speak of manure. Varahamihira in the Brhat Samhita[1] says,. "To promote inflorescence and fructification, a mixture of one adhaka (64 palas) of sesame, 2 adhakas of excreta of goats or sheep, one prastha (16 palas) of barely powder, one tula of beef thrown into one drona (256 palas) of water and standing over for seven nights should be poured round the roots of the plant" (17, 18). He further prescribes that the seeds before sowing should be treated as follows: —They should be taken up in the palm greased with ghee and thrown into milk; on the following day the seeds should be takenout of the milk with greased fingers and the mass separated into single seeds. This process is to be repeated for 10 successive days. Then, the seeds are to be carefully rubbed with cow-dung and steamed in a vessel containing pork or venison. Then the seeds are to be sown with the above mentioned flesh and lard in a soil where previously sesame was sown and dug up, or trodden down and then sprinkled daily with water mixed with Ksira. (19, 20)

"To ensure the growth of *Ballaris* (i.e., sprouting and the growth of luxurious stem and foliage), the seeds should be properly soaked in an infusion of powdered paddy, masa (bean), sesame and barley mcomposing flesh and then steamed with *Haridra* (turmeric). This process will succeed even with *Tintith* (tamarindus Indica). For the *Kapittha* (Feronia elephantum) the seeds should be soaked for about 2 minutes (literally, such length of time as it would take one to make a hundred rhythmic claps with the palms (talsudah)in a decoction of 8 roots,*Asphota* (Jamine), *Amalaki* (Phyllanthus umbilicus), *Vetula* (calamus rotung), *Suryavalti* (Gynandropsis pentaphyla), *Syama* (Echites fructescens) and *Atimuktaka* (Aganosma caryophyllata) boiled in milk. The seeds then should be dried in the sun. This process should be repeated for 30 days. A circular hole should be dug in the ground, a cubit in diameter and 2 cubic deep, and this should be filled with the milky decoction. When the hole dries up, it should be burnt with fire and then pasted over with ashes mixed with ghee and honey. Three inches of soil should now be thrown in, then the powder of bean, sesame and barley, and then again three inches of soil. Finally washings of fish should be sprinkled and the mud should be beaten and reduced to a thick consistency; then the seeds previously prepared should be placed in the hole under three inches of the soil and fish washings (with fish). This will lead to luxuriant ramification and foliage which will excite wonder.[1] 21-26).

The Agnipurana[2] gives the following directions:

"A tree becomes laden with flowers and fruits by manuring the soil with powdered barley, sesamum and the offal matter of a goat mixed together and soaked in washings of beef for seven consecutive nights. A good growth of trees is secured by sprinkling them with the washings of fish." While Khana advises the cultivators thus:

"O worthy cultivator, for a vigorous growth of bamboo, give an infusion of powdered paddy to its roots, for the growth of Arum, ash, and for that of coconuts, salt.

In the Arthasastra it is stated that:

"The seeds of grains are to be exposed to mist and heat *(tusara-payanam usnam ca)* for seven nights; the seeds of Kosi such as Mudga, masa etc, are treated similarly for 3 nights; the shoots of sugarcane and the like (*kandabija*) are plastered at the cut with the mixture of honey, clarified butter, the fat of hogs an cow-dung, the bulbous roots (*kanda*) with honey and clarified butter, cotton-seeds with cowdung; and water pits at the root of trees are to be burnt and manured with bones and cow-dung at proper sea sons. The sprouts of seeds when grown are to be manured with a fresh haul of very small fish and irrigated with the milk of snuhi (Euphorbia antiquorum)".

## V

## COLLECTION AND TREATMENT OF SEEDS

About the collection and treatment of seeds. Parasara says,—

All kinds of seed should be collected n the month of *Magha or Phalguna*. They are then to be well dried in the sun and exposed to dew at night". "The *Putikas* (small receptables) are afterwards made and the seeds kept in them". "Different kinds of seed must be kept separate, for mixed seeds are not good." "After the seeds have been placed, in the *Putika* must be well closed; grass that may grow out of it must be up-rooted or when the seeds sprout, the field will be full of grass."[2]

## VI

## CONSTRUCTION OF AGRICULTURAL IMPLEMENTS

Parasara describes the plough and its accessories thus: "The plough consists essentially of the following 8 parts: Isa (the pole of the plough), Yuga (the yoke), Niryola (the rod of the plough exclusive of the pole and the share), Niryolapasika (iron plates that fix the share to the Niryola). (There are two pairs of Pasika), Halasthaanu (a strong piece of wood that is fixed to the Niryola at the end opposite to which the plough-share is fixed; this is held by the cultivator while ploughing the field), Addacalla (the pins of the yoke where the bullocks are tied), Saula (an extra piece of wood that tightly fixes the Niryola to the pole) and Paccani (goad)".

"*Isa* is 5 cubits long, sthanu 2 1/2 cubits, niryola 1 1/2 cubits, *yuga karyasamanak* (?), *niryola-pasika* and *addacalla dvadasasukhmone* i.e., about 9 inches, taking the breadth of a finger to be approximately 3/4 inch) and *saula* an Arbi (nearly a cubit). Pacani is made of bamboo with iron-top and about 3 or 41/2 feet long, (taking the transverse length of a fist (Musthi) to be approximately 4 inches."

"*Abandha* (a rod of iron which prevents the *niryola* from getting out of the pole) must be cyclindrical and (pasudsadva) about one foot long, *Yoktra* (the tie of the yoke) 4 cubits in length, a rope 5 cubits long and Phala (ploughshare) I cubit or (pasushuladviko haste) I cubit and 4 inches, Pasika is Navasula (nearly 7 inches) and looks like a laf of Arka (calotropis gigantea). *Viddhaka* a big hoeing instrument) ha twenty-one spikes and the harrow is 9 cubits long."

Besides the plough and its accessories as enumerated by Parasara, there were also the following agricultural implements:

*Srni* (sickle), *Khanitra* (hoe), *Musala* (pestle), *Udukhala* (mortar), Calani (sieve), *Sthivi* (granavry), *Methi* (the post of the threshing floor round which cattle turn to thresh out the grains), etc.

## VII

## PLOUGHING THE LAND

"In order that the crops may have a luxuriant growth", Khana says, "the cultivator should plough the lan under the auspices of the asterisms known as, *Svati, Uttaraphalgam, Uttarasadha, Uttarabhadrapada, Mrgasira, Mula, Punarvasu, Pusya, Srvana or Hasta* and on Fridays, Mondays and Wednesdays."It is unwise to begin ploughin on the day of the full moon or the new moon", "the land should be ploughed 16 times for radish, half the number of times for cotton, half of that for paddy and none for betel." "The soil for radish must be as soft as cotton and for sugarcane, the soil must be ploughed to dust."

## VIII

## SOWING, PLANTING, TREATMENT OF PLANTS, ETC.

"Aman paddy, jute, ginger, arums, turmeric, mango, pumpkin, gourd, cucumber etc. are to be sown in the month of Vaisakha; and sugarcane, plantain and betel are also to be planted in this month if they have not been planted in *Caitra.*" "The field is manured generally towards the end of *Jyaistha.* In this month Jyaisthik paddy ripens." "The month of Asadha is the best for planting autumn paddy (i.e., for transplanting the aman paddy that is sown in Vaisakha), and also for planting mango, lichis, coconut, flower plants, betel etc. In this month, Aus paddy begins to ripen." "Pepper, tobacco, kalai and kaluttha (doclichos biflorus) are to be sown in Sravana. Old trees will surely bear fruits if the solid at their rots is turned up enthuse month and fresh soil dumped there in the month of *Agrahayana.*" 'Turnip, sesa mum, mudga (mango) and pepper are to be sown towards the end of *Bhadra.* IN this month, Aus paddy fully ripens; an dhoeing is done and water drained off from the field (in which aman pady had been sown) leaving just enough for only the roots tobe under water.

"In the month of Asvina, the ground must be prepared for the winter crops. Godhuma (wheat), mustard, kalai, cabbage, potato, radish, beet etc. are to be sown in this month and arrangements made to preserve water in paddy-fields." "Barley, pes, coriander, water-melon, cucumber and gourd are to be sown in the month of Kartika." "In Agrahayana, aman paddy ripens and pumpkin is sown." "In Phalguna, winter crops ripen, and if it rains sufficiently, aus paddy is sown."

Now-a-days, Aus is mostly sown in the month of vaisakha. In *Phalguna, Jyaishik paddy* is sown. There is usually a scarcity of rain in the months of Phalguna and Caitra. That is why *Jyaaisthik* paddy requires a marshy land for its growth.

**Rotation of crops.**

Rotation of crops was known early in India. *Aus paddy* and winter crops are thus sown in the same field by rotation—the former in spring and the latter in Autumn.

"Khana says, "my worthy cultivator, plant radish towards the end of the 3rd season of the year, sow mustard towards the end of Autumn. And if you mind to make money, sow maze in the following month of *Caitra*."

In the Kautiliya Arthasastra (Eng. Trans. 2nd ed., pp. 139 f.) the directions are given about sowing.

*Sali* (a kind of (rice), *vrihi* (rice), *kodrava* (paspalum scorbiculatum), trial (sesamum), *priyangu* (panic seed) *daraka* (?) and *varaka* (phaseolus troubles) are to be sown at the commencement (purvavapa) of the rainy season. Mudga (phaseolus mango), *masa* (ph. radiatus) and saibya (?) are to be sown in the middle of the season. *Kusumbha* (safflower), *masura* (Ervum hirsutum), *kulattha* (dolichos biflorus) *yava* (barley), *godhuma* (wheat) *kalaya* (leguminous seeds), atasi (linseed) and *sarsapa* (mustard) are to be sown last".

The following directions are given in the Brhatsamhita (chap. 55) and the Agni-purana about the planting of trees:

*Brhatsamhita:* "It is best to plant trees at intervals of 20 cubits, next at 16, and 12 cubits interval is the minimum that can be prescribed." 12.

*Agni-purana*: "Trees planted in rows twenty cubits apart should be deemed as the best planted, while those having an intervening space of 16 or 10 cubits between them should be deemed as ranking second best in respect of fruitfulness. Trees should be transplanted 12 times, they should not be planted too close or adjoining each other, in which case, the breaches that touch one another should be lopped off with a chopper, as otherwise, they would bear no fruit. Should barrenness be apprehended, the leaves and branches of a tree ought be sprinkled over with a solution of cold water saturated with Vidanga, clarified butter, pulverised Mudga, Masa and Kaluttha pulse. Similarly, a tree sprinkled with cold water and clarified butter becomes laden with abundant fruits and blossoms.

The method of propagation by cuttings and graftings was also known to India from very early times. The following lines from the Brhatsamhita (chapter 55) will be illustrative: "*Kanthal* (Jack-fruit tree), *Asoka, Kadati* (plantain), *Jambum Lakuca, Dadimba, Draksa, Palibata, Bijapura* (Matulanga) *Atimukta ka*—these are the plants to be propagated by means of cuttings besmeared with cow-dung" (4-5). "Better than this method is the method of propagation by graftings. This can be down in two ways,—the cuttings of one plant is either inserted in the root of another plant or on the stem of another plant. Grafts should be smeared with cow-dung. For transplanting the plants should be smeared from root to the top with ghee (clarified butter), sesame oil, honey of the ksudra variety of bees of the Usira. (Andropogon Langiger or Andropogon Chitrarum), the Vidanga (Embelica robes) milk and cow-dung." (6-7) "The most suitable ground for planting is soft soil that has been sown with sesamum indicum and dug up and trodden with sesame in flower."

Directions for sowing paddy are elaborately given by Parasara in his *Krsi-samgraha* thus:—

"Sowing in the month of aisakha is best, in the month of/yaistha is tolerable, Asadha bad and Sravana worth. Transplanting should best be done in Asadha. It is bad to do it in sravana and worst in Bhadra.

"After the seeds have been sown, the field is to be harrowed, otherwise the seeds are not distributed uniformly over the field." 169.

"The seeds are of two kinds—one for sowing and the other for planting. Only healthy seeds are used for sowing. Unhealthy seeds are for planting (i.e., the seeds are sown and when they sprout, the shoots are transplanted). Full-grown plant should not be raised; transplanting is to be done while the plants are young. If planting is done in the month of Srvana, there must remain one cubit of ground between any two plants; if in Bhadra, 1/2 cubit, and if the Asvina, (i.e., about 3 inches). (170, 171, 172).

### On Second Ploughing

*Parasara* says, " We have already said that the seeds are of two kinds. In the case of those for which transplanting is not necessary, the field is ploughed for a second time in the month of *Asadha* or *Sravana*, after the seeds grow into plants. This destroys some of them but they subsequently turn into manure and give the rest a healthier growth. "For want of sufficient timely rains, the second ploughing may be done in the month of *Bhadra*, but in that case, the produce will only be half. If the second ploughing is not done before Asvina, the prospect of harvest is little." (174). "Planting or the second ploughing should not be done in marshy land neither should manure be given; only the grass is to be cleared out." (175).

"If after planting, the field is not hoed, the crops can't grow in abundance, nor yield a good harvest. If hoeing is done in the month of Sravana or *Bhadra*, the harvest is doubled thereby, even if grass may again grow. If another hoeing is done in the month of *Asvina*, corns grow as plentifully as Masa." (176-178).

### Drainage and Provision for Water

"So that the paddy may not get diseased (Neruja), the water in the field is to be drained off in the month of Bhadra, leaving just enough for only the roots to be under water." (180)

"The full who does not make arrangements in Asvina and Kartika to preserve water should not expect a harvest." (183).

## IX

## REAPING AND STORING

### Harvesting

"In the month of *Pausa* (after the *Pusya-yatra*), the wise cultivator reaps the full harvest and after thrashing the corns, measure the grants with ADHAKA. The measured grains are then stored and are never spent in Pausa even erroneously." (217, 218).

*Khana says:* "The corns ripen 30 days after the appearance of flowers and 13 days after the ears bend low." "the corns ripen in the month of *Agrahayana.* They should be reaped in Pausa threshed in Magha and husked in Phalguna."

## CONCLUSION

Thus beginning at the time of the Rg-veda, the science and method of cultivation steadily developed and attained a high stage of perfection by the 4th century B.C. In the Vedic age, agriculture was the general occupation; each family possessed a number of corn fields in well-marked holdings measured off according to the standard of measurement prevailing in those days. The occupation was then considered noble and each respectable householder was eager to possess a number of cattle and fertile corn-fields. By the time of the Yajurveda, caste-system became fully established and agriculture became the occupation of Vaisyas. In the age of the Brahmanas we find that industrial workers were gradually sinking in estimation, yet agriculture being the staple industry of th country went on developing in the Sutra and the Epic periods. After the disruption of the Maurya empire, the political history of India became full of vicissitudes and under inimical political circumstances, agriculture instead of being a concern of the state and a matter of expert knowledge, was relegated to the lowest strata of population. The wisdom that grew in the course of ages remained buried in literature not easily accessible to the mas. Thus neglected and uncared for, the art of cultivation gradually decayed in India.[2]

*R. Ganguli*

## REFERENCES

1. *The Ancient Punjab.*
2. *Vedic Index,* II, 348.
3. *Rgveda,* X, 101, 8 (Wilson).
4. *Rv.,* X, 101, 3, 5. (Wilson).
5. Macdonell and Keith, *Vedic Index,* 1, 139.
6. Vode *Gupta Press Panjika.*
7. *Arthasastra,* English translation by Dr. Shamasastry, p. 108.
8. *Krsi-samgrahah* (Bangabashi Edition). All the verses from Krisisamgrahah are translated for hefirst time bythe author.
9. Avarta, Samuvrta, Puskara and Drona are the four types into which clouds are divided.
10. *Adhak* is a unit of measure in Astonomy. It is equal to 100 jojans square and 30 ypjanas high (1 yojana=about 7 miles).

11. Vide *Krsi-samgrahah,*
12. Vide *Krsi-samgrahah.*
13. Vide *Krsi-samgrahah.*
14. VIde *Krsi-samgrahah.*
15. *Loc, cit.,* p. 139.
16. *Loc. cit.,* p. 139.
17. Vide, *Krsi-parasarah.*
18. Vide *Krsi-Samgrahah.*
19. Vedic *Index, ii,* 146. .
20. *Brhatsamhita* edited by Dr. Kern, Chaptr 55, p. 304.
21. Translation of these verses from the Brhatsmahita are taken from Dr. Seal's Positive Sciences of the Hindus.
22. Translation by M.N. Dutt, vol. II, p. 1038.
23. Shyama Sastri's translation, 2nd ed., p. 141.
24. *Krsi Samgrahah,* 148-151.
25. *Loc. cit.,* verses 110-117 and 96, 97.
26. Here Adhaka is a unit of measure for grains. It is vessel having a volume of 216 cubic inches.
27. I am thankful to Dr. B.M. Barua of the Calcutta University for many suggestions.

# 5

# Rudra-Siva—as an Agriculture Deity

The great god Siva appears in an interesting light in medieval Bengali literature. He is represented as a cultivator who ploughs his fields, sows seeds, takes out weeds, cuts grass and binding it into a sheaf, carries it home on his head. While he thus works hard on his land like a humble farm lad we should not suppose that he has forgotten his own divine origin. At his command Visvakarma, the divine architect, hastens to prepare a golden plough and a golden sickle for his use and Indra, the mighty king of heaven, hastens to pour down water from the clouds to irrigate his lands. The poets are careful to explain why Siva has to take to agriculture. He is rather an impecunious god. In truth, mendicancy is his only means of subsistence. The miserable condition of his household is further aggravated by his fondness for drugs and low company. Domestic quarrels naturally break out often, in these circumstances, and in these quarrels the poor old god is worsted by his young wife. His misery arouses deep pity in the heart of a devotee who requests the god to take to agriculture,—to sow paddy so that he may have food, and cotton so that he may not have to go naked or wear skins for want of cloth.[1] There are different accounts. In another account it is Parvati, Siva's wife, who opersuades her husband to take to agriculture and sends him to Indra for grant of a suitable land for cultivation. After he has received from Indra such lands as he wanted he is in difficulty about implements and bullocks. Visva-karma makes him a plough and a goad from his trident and Parati advises him to yoke his own ox and her vehicle, tiger, to the plough. This he does and ploughs his fields with the help of Bhima and sows paddy etc.[2]

Siva is also spoken of in some part of this literature as the garden-keeper of Dharma.[3] In the half-Buddhistic and half-Sivaite Gambhira and Gajana festivals which appear to have originated at an early date and must have preceded Ramai Pandita's work, there are references to the agricultural role of Siva. On the day of Ahara puja, a feature common to both Gambhira and gajana Festivals, cultivation by Siva is represented by the *bhaktas* or devotees before the spectators. In the Vandana or salutation ceremony verses are recited containing references to cultivation by Siva.[4] In the later Bengali literature this role of Siva disappears. Curiously enough,an elaborate account of cultivation by Siva appears in the work of Ramesvara Bhattacarya who lived as late as the first half of the 18th century. The language of this poet is archaic and it seems that his work is in a large measure, a collection of old folk songs about Siva. In the same literature in which Siva appears as a cultivator there is another feature in the representation of Siva to which reference has been made and which gradually rose into prominence as the conception of cultivator Siva faded. He is a typical vagabond god as well as a god of vagabonds. One poet calls him *Badiyar Po* i.e. son of a Bediya, a well-known, wandering gypsy tribe in India. He is also called *Bhangara* (i.e. addicted to *bhang), Baul* (i.e. a wandering mendicant singer), *Pagala* (i.e. crazy).[5] He keeps low company and his partiality for koch women is notorious.

From the above we find that Siva appears mainly n two aspects in the medieval Bengali literature and folk-songs and festivals—a cultivator god and a mendicant, crazy, a god of the lowly. Of these two conceptions, that of the cultivator god is the earlier. We propose in the present essay to enquire into the origin of this cultivator god of the lowly rustics and for this purpose we shall examine, first, some instances of th worship of Siva in Brahmanical society and then in non-Brahmanical society that is, among the semi-Hinduised tribes, and tribal peoples and next examine some of the references to Rudra-Siva in the early literature. It should be added that we do not propose to consider sectarian worship of Siva, that is, his esoteric worship by the Saiva sects.

### In Brahmanical Society

Non-sectarian worship is offered to Siva as a household or family god, an ista-devata (village deity). Apart from general worship of Siva in any of these forms special worship is also offered to him in Bengal and other provinces for such specific purposes as obtaining of offspring and cure of diseases, mainly chronic, while special worship of Siva by young unmarried girls for obtaining suitable husbands appears to be confined to Bengal. The worship of a local form of Siva known as Penco connected with infantile disease is existent in Bengal. The only instance of the worship of Siva which may be brought into indirect connection with agriculture is reported from Sylhet where siva is worshipped in the cattle-shed for the purpose of protection of cattle.[6]

### Siva-worship among semi-Hinduised Tribes and Tribal peoples

Among Jashpur Kisans each village has two or more sacred groves one of which is sacred to Mahdeo who is specially invoked at the festival of harvest home.[7] An effigy of a man is sacrificed annually before Mahadeo by the tribes living near Lohardaga for rain and plentiful harvest. A similar offering is made by the Gonds to Baradeo.[8] Baradeo appears to be variant of Mahadeo. On the occasion of the Arwa or Kharway, that is, harvest home festival, the Binjhias offer as sacrifice to Mahadeo, the Baiga officiating.[9] Among the Biyars two stone pillars representing Dharti-Mahadeva are propitiated when harvest is gathered, with burnt offerings. Vows are also made to them by people in trouble.[10] To Bhairo and seven sisters who scatter cholera, small-poxm, cattle-plague, etc.

a rude shrine is erected by the Rautias in the centre of every village consisting of a raised plinth covered by a thatched or tiled roof. The god and his sisters are represented by small mounds of earth.[11] The Rajputana Bhils worship Baba Deo who has his seat at Deogarh Bariya in Rewa Kantha Agency at the time of harvest home.[12] The function of Darwa Gosain worshipped by the Mals of Rajmahal is that of the guardian deity of village or its gate-keeper.[13] He seems to be allied to Darha of the Oraons who is connected with agriculture.[14] Birnatha is worshipped by the Ahirs as a protector of cattle. Occasional views are also made to him in sickness and distress. Worship is worshipped by the Ahirs as a protector of cattle. Occasional vows are also made to him in sickness and distress. Worship is offered in the morning when the cattle are sent to jungle in hot weather.[15] Crooke is inclined to think that this cult has been borrowed from the Kols. But the fact that th cult obtains among the Ahirs who are Hindus and among Hinduised tribes as also the fact that Birnatha is regarded as a protector of cattle tend to show that he is probably a local form of Mahadeva.

The name Nattan Kal is given in the Tamil country to a post set up and worshipped in a nuptial booth as well as small stone set up at the entrance of a village. Bishop White head is of opinion that the Nattan Kal set up in wedding booth may represent the spirit who presides over the procreation of children and may possibly be a phallic emblem.[16] Iyyappan or Sastha according to Mr. Iyer[17] is believed to be the supreme god among th non-Aryan aborigines in South India. He is the chief of the *bhutas* and is identified in this respect with Siva. He is that favourite god of the lower castes in the rural parts in Malabar, Travancore and Cochin and is represented as riding with a sword in hand, clearing the country of all obnoxious spirits. According to the popular legend of origin, sastha is the result of the union of siva and Visnu in the for of Mohini. Mr. Iyer thinks that this is a piece of invention of th Brahmin priesthood to destroy the old idea of non-Aryan gods which still influences the masses. "As people especially the Sudras and lower castes were largely agricultural and fond of hunting and they had to live and work in forests they began the propitiation of this deity for protection against demons and illness." This Iyyappan is also known as Iyenar or ayenarappan in the Tamil country.[18] He stands in the open under a tree with clay images of horses, elephants, dogs and warriors (viras). IN Vandipaliam, Cuddalore, when animals and fowls are sacrificed, a bottle of arrack, a pot of toddy, some cheroots, some ganja and opium and dried first are presented to the god. After recovery from sickness or to commemorate any piece of good fortune the villagers place fresh clay horses round his shrine. Every year after harvest time a festival is held in his honour when animals are sacrificed and the images of the god the drawn about the village streets on rude cay horses.[19] A deity called Muni is worshipped by the Eravallens in Cochin in order to protect cattle and to have a good harvest. "Kali and Muni are worshipped in forests. Offerings are made to Kali and Muni when the tribe plough and sow and reap. They protect corn from destruction by wild beasts.[20] The same god is also worshipped by the Pulluvans and by the Kaniyanas for protection of cattle.[21] From his association with Kali and connection with agriculture and cattle as well as from his name it may be reasonably inferred that mani is a variant of Siva.

Thus, Mahadeva and various other local gods whom it is possible to affiliate to him either on account of their names or similarity of functions or association with certain deity or deities are connected with agriculture, cure of illness and protection of cattle. They are also connected, though in a lesser degree, with protection from wild beasts and the function of a guardian deity of village or its gatekeeper. The clearly local character of these gods and their worships is shown by the animal sacrifices (including vicarious human sacrifice) offered to them and by the fact that the officiating priest is invariably a tribesman.

## In the Rgveda

We shall have now to direct our investigation into the ancient literature. The Rgvedic Rudra is a fierce, wrathful, malevolent god.[22] He is a destroyer and is connected with death.[23] He is also the best among physicians and distributes medicines.[24] He is Pasupa or protector of cattle.[25]

## In the Later Vedas and Vedic Literature

Rudra is clothed in skins and dwells in mountains.[26] He is the lord of pilferers and robbers.[27] He is called Pasupati.[28] He is connected with the Vratyas or outcastes, serpents, bids of evil omens and howling dogs.[29] Direct mention of his name is to be avoided.[30] Offerings are prescribed to Rudra for the sake of cattle.[31] Rudra's hosts the ganas attack men with death and disease and to appease them the bloody entrails of victims are offered in sacrifice.[32] Obeisance and prayer to Rudra are prescribed for safety in the course of journey by a man and also when a man comes to a place where four roads meet or crosses a river or enters a frost.[33] the sacrifice of a bull to Rudra to appease him (him (the Sulgava sacrifice is to be performed in a cattle-shed for averting cattle disease. Rudra dwells among cattle as well as among serpents.[35] Rudra is thus a terrible god having connection with wilderness, cemeteries, mountains, old trees etc., i.e., with dangerous places and inauspicious things and with thieves, outcastes, Nisadas, etc. In the Upanisads the conception of Rudra-Siva is provided with a philosophical basis. Rudra-Sivas's connection with yoga appears in some of the later Upanisads, e.g., *Atharvasitras, Svetasvetara.*

To sum up: In the Rgveda rudra is a malevolent god of death and destruction and at the same time a benevolent physician who cures diseases with his healing remedies. He is also connected with cattle, though this connection is no t prominent. In the later Vedic literature his malevolence increases, his connection with cattle receives prominence and though his connection with diseases is maintained his healing power is almost forgotten. His connection with forests and trees receives some attention. He deposited his weapons on trees. When a man came near an old tree specific formulae in his honour had to be recited. It has been suggested[35a] that the suspension on trees of the offerings at the Tryambak offering of the Srauta ritual is probably a vegetation or tree ritual.[36] He also appears in the midst of new surroundings, such as, hills, forests, deserted places, outcastes, thieves, vagabonds, savges clothed in animal skins, serpents, demons, etc. The picture is that of a sombre, malevolent, inauspicious, disreputable god.

**In the Epic**

In the Mahabharata the more popular names of Rudra are Siva, Mahadeva, Sankara, Pasupati, etc. he is as of old, connected with death and disease. The battle-ground of deth is his playground.[37] His wrath produces fever.[38] He is connected with serpents and his ornaments are Naas worn as earrings and girdle, his clothes are of snake-skin. He dispels fear of snakes. His connection with cattle continues.[39] The bull, his favourite animal in Sulagava sacrifice, becomes his vehicle. Low company still delights him. When Arjuna worshipped Pasupati in the Himalayas for obtaining the Pasupat weapon he appeared before Arjuna in the form of a Kirata. His outlandish characteristics are recognized when he is given such names as Kumbhakarna, Sankukarna Gokarna, etc.[40] He is Ganakarta, Ganapati,[41] He wears tiger-skin.[42] He is Nisacara, Pretacari, Bhutacari.[43] He is almost a raksasa having a distorted face, sword-like tongue and taste for raw flesh. In fact, he is greedy of flesh as a jackal.[44] Being a god of the outcastes the god had himself lost caste among the gods. when asked why he had no invited Mahesvara to his sacrifice Daksa said,—I know the eleven Rudras but I do not know who Mahesvara is.[45] He tells his consort that following the old custom, the gods did not assign him any share in sacrifice.[46]

Manifold are the new attributes that are given him. He is a sort of incantation god[47] fond of dancing and music.[48] He is an ascetic and a Yogi.[49] He is a giver of offspring. Thus he grants Krsna the boon of a worthy son. To Anasuya, wife of Atri, he granted the boon of a son.[50] The most important new fact in connection with Rudra-Siva in the epic is his worship in the phallic emblem.

The epic gives him several epithets which are indicative of his association with trees. Thus he is described as *Nyagrodharupo nyagrodha* etc. Again *Vrksakaro vrksaketu* etc.[51] He is gavampati as well as *Vrksanam pati and Vanaspatinam pati.*[52] As a forest-god it is natural enough that these epithets should be given to rudra Siva. But there are other epithets which show him in the light of a giver of food. He is the creator of food, giver of food, lord of food.[53] As a giver of food he comes to be connected with the process by which food is produced. He takes care of the seed and the field and is th creator of both *Vijaksetrabhipalaya* etc.[54] His connection with gegetationis clearly brought out when it is stated that the plants (ausadhi) were produced from him.[55]

Nearly all the old attributes of Rudra remain to the epic Rudra-Siva but they pale into insignificance before the multiplicity of the new attributes that are given him; the new god is vastly more important than the old god. Among his more important new attributes are his power to grant offspring, his asceticism and yoga, his assumption of th linga-form and his connection with vegetation; among his old attributes which are not forgotten but which are relegated to the background are his connection with cattle and his healing power. Rudra-Siva is no longer an inauspicious,injurious god, he is siva i.e. auspicious, Mahadeva and Mahesvara, the great god.

### In the Puranas

No notice is taken in the Puranas of this connection of Rudra-Siva with vegetation while the other aspects of his character are accentuated with this difference that his fearful attributes are transferred to his consort who already comes into prominence in the epic. The food-giving attribute is also taken away from him and transferred to his consort as a virtue more appropriately adorning a female. In fact, the emergence of the worship of the *Linga* into prominence in the epic so effectually cast into shade other affiliations of Siva that it is not surprising that his unimportant connection with agriculture should be lost sight of in the Puranas.

### Conclusion

In the medieval Bengali literature Siva is, as we have seen, a cultivator-god, but he is not a vague, impersonal lord of fields and seeds, of forests and trees and of food but a real cultivator driving the plough, turning up the soil, sowing seeds, weeding fields, returning home with a head load of grass after a day's work. He is not a divinity worshipped from distance for increasing the fertility of the sailor prayed to grant bumper crops or keep away pests but a personal, anthropomorphic god actively engaged in agriculture. Such a conception of god is quite early, certainly pre-Epic and apparently pre-vedic. As this cultivator-Siva first comes to notice in a literature of which the earliest date cannot be earlier than the 10th century, it must be clearly assumed that the worship of this god prevailed among a people who had no use for vague, impersonal gods, who did not understand gods who were removed from their everyday concerns. That such was the fact is proved by the evidence of medieval and later Bengali literature. He appears definitely as a god of rustics, vagabons, low-caste people and despised outcastes living on the fringe of respectable society. He recalls associations of vagabond gypsy, mendicants, crazy minstrels, half-savage tribes etc. There are, thus, two aspects of the medieval Bengal Siva, —he was a rustic, cultivator-god and a half-mad mendicant god of the lower folk. It is this half-mad mendicant god of the Lowe folk. It is this half-mad mendicant god taking active interest in agriculture that hold the imagination of rural Bengal and is sung in folk-songs. He is the god of all, high and low. No barrier of caste restrictions exists in his worship.

The malevolent, destroying rudra and the benevolent, healing Rudra of the *Rgveda*, apparently a syncretism of different deities cannot probably be the prototype of this rustic cultivator Siva. The vagabond god of the hills and forests and of the Vratyas and skin-clothed savages of the later Vedas has some features in common with him. Rudra has, however, pronounced inauspicious, sinister associations and demoniacal features in the later Vedic literature and this aspect of Rudra can be traced to the *Rgveda*. The vagabond Rudra is not only a god later than the sinister Rudra but also different from him. An important point which requires to be noted is that Rudra's connection with vegetation appears in the sutra period. The evidence of both, *srauta* and *Grbya-Sutras* proves that this aspect of rudra was regarded as of some importance. In the epic this connection with vegetation is further developed. From being the lord of forests and trees, origin of plants, creepers, grass and giver of fruits and flowers he comes to taking care of fields and seeds.Next, he is conceived of as the lord and giver of food. The sectarian spirit of the staunch Sivaite is already apparent in the epic in the discourses of the sage Upamanyu. It is, however, curious to note that even a fiery propagator of linga-worship like Upamanyu who initiated Krsna into his favourite cult should ask from Mahadeva such as boon as rice and milk in plenty. As we do not find any pre-epic connection of Rudra with agriculture, though connection with vegetation or vegetation-spirit may be inferred, it is a plausible hypothesis that there was a local god of agriculture who syncretised with the epic Mahadeva with a considerable loss of local features and that it is in this unknown local god that we should try to find the cultivator-Siva of medieval Bengal. He has, however, other pronounced features which lend strong support to the view that the cultivator-Siva in both his aspects, that is, a cultivator and a mendicant, half-mad god, may have been derived from the epic Mahadeva. It has been seen that the cultivator-god of medieval Bengal is also a god of the lowly,a mendicant, crazy god who dances and sings. The epic, gives Mahadeva such epithets as *unmadab* (crazy),[56] clothed in rags[59] etc. These are curious epithets for a god who is identified with the highest spirit of the yogis and who grants boons to and orders all other goods including Krsna and Brahma! This mendicant crazy Mahadeva of the epic, essentially a god of the lowly rustics was, it would appear, the source or copy of the medieval Bengal Siva in on of the two forms, namely, the crazy, mendicant god of the lowly and the rustic cultivator god. This wandering, crazy god of the lowly appears to have an unbroken tradition from the times of the *Yajur Veda* and the *Atarva Veda.* This tradition was embodied in Rudra in pre-epic times in Siva in Epic and Puranic times and in Bhola Mahesvara or Budo-Siva in medieval Bengal.In the non-sectarian special worship of Siva in Brahmanical society we find no trace of this cultivator, mendicant god. Among semi-Hinduised tribes and tribal peoples, is worshipped a number of local gods whom we have affiliated to Siva for the following reasons:—

(1) On account of their names which may be brought into connection with one or other of the various names under which Siva is known.

(2) Similarity of their functions with the functions attributed to Siva, and

(3) Association with deities having known connection with Siva.

On an examination of their various attributes the following results are obtained:

(1) Connection with agriculture

(2) Cure of maladies

(3) Protection of cattle

(4) Removal of such pests as wild beasts which attack men and cattle and destroy crops.

(5) Keeping watch over a village as its guardian deity or gate-keeper.

Of these, curing of maladies and protection of cattle are earl y traits; removal of pests is connected partly with agricultural aspect and partly with the aspect of a guardian deity. Of all the attributes mentioned above connection with agriculture and diseases is more in evidence. Offerings are made to the god at harvest, and animals and such birds as fowls, pigeons etc. are sacrificed. In two instances, human effigies are sacrificed, clearly a substitute for actual human sacrifice. The purpose of such sacrifice is to get seasonal rains and rich crops. Animal sacrifices were offered to Rudra in the Sutra period not for crops but for prevention of cattle-plague, and raw flesh-eating is a trait of epic Rudra-Siva who as a Kapali also appears to be connected with human sacrifice. It is obvious, however, that the Mahadeva worshipped at harvest with animal sacrifices cannot be the rustic cultivator-god; his conception indicates a more developed notion of the deity, more akin to the epic lord of food who takes care of fields and seeds.

We are inclined, therefore, to the view that the cultivator-Siva may have been partly derived from a local god of agriculture and food who syncretised with the epic Mahadeva, but it is more probable that this aspect of the epic god was a development of the connection with vegetation which he inherited from the vedic Rudra. We are, further, inclined to the view that the other aspect of the cultivator-Siva namely, the aspect of caxy, mendicant god of the lowly, is also derived from the old Vedic Rudra. This being the more prominent aspect, embodying an old. uninterrupted tradition, we are persuaded that in the medieval Bengal Siva we see the old, wandering god of the outcaste and semi-saavage tribes of the Satarudriya humans of the *Atharva Veda* and the Mahadeva of the epic; his intensely realistic agricultural aspect was probably a local development under the special conditions of medieval, rural Bengal, the basic tradition having been derived from the Vedic Rudra connected with vegetation and the epic Siva connected with vegetation, food and the process of production of food.

## REFERENCES

1. *Sunya Purana* by Ramai Pandita, D.C. Sen, Vanga Sahitya Paricaya (Cal. University), pt. I, pp. 112 ff.

2. *Dharma-Puja Vidhana* by Ramai Pandita, edited by N.G. Banerjee (Vangiya Sahitya parisat) pp. 227ff.

3. *Dharma-Puja Vidhana* by Ramai Pandita, edited by N.G. Banerjee (Vangiya Sahitya Parisat) pp. 227ff.

4. See *Adyer Gambira* by Hridas Palit (Jatiya Siksa Samiti, Malda) and Folk Elements in Hindu Culture by Benoy K. Sarkar, pp. 57, 90.

5. IAS, vol. 1,3, pp. 181 f.; D.C. Sen, *Vanga Sahitya Paricaya.*

6. Folk Customs and folklore of Sylher District by P. Bhattacharya (Man in India, vol. X, 2-4, p. 261).

7. Dalton, *Descriptive Ethnology of Bengal*, p. 132.

8. Dalton, *Descriptive Ethnology of Bengal*, p. 259.

9. Risley, *Tribes and Castes of Bengal*, vol. I, p. 136.

10. Crooke, *Tribes and Caste of North-Western Provinces and Oudh*, vol. 2, pp. 136-37.

11. Risley, *Op.cit.*, vol. 2, p. 203.

12. Imperial Gazetteer, Provincial Series—Rajputana, p. 88.

13. Risley, *Op.cit.*, vol. 2, p. 58.

14. Dalton, *Op. cit.*, p. 268.

15. Crooke, *Op. cit.*, vol. 1, p. 63.

16. Whitehead, The Village Gods of south India, pp. 41, 42.

17. L.K. Anantha Krishna Iyer, The Cochin Tribes and Castes, pp. 1, 312-313.

18. Whitchead, *Op.cit.*, p. 90.

19. Monir Williams, Religious Thoughts and Life in India, pt. I, pp. 220 f.

20. Iyer, *Op. cit.*, p. 47.

21. Iyer, *Op. cit.*, p. 223.

22. *Rig Veda*, 11, 33 9, 11.

23. Rv. 1. 114 7, 8, 10.

24. Rv. vii. 46. 2.; 11. 33. 2, 12.

25. Rv. 1, 114. 9.

26. Vaj. Samhita, 3, 61.

27. Vs. 16, 18, 28.

28. Vs. 39, 8.

29. *Atharva Veda*, xvii, liii, 26, 27; xi, 2, 2, 2, 11; xi, 2, 30.

30. Ait, Br., ii, 34, 7.

31. *Pancavin*, Brahmana, vii. 9. 18.

32. *Grhya Sutra Sankh*. 4. 19.

33. *PGS*. 11. 17. 7 f.; HGS. I. 18. 8 f.

34. *AGS*. IV. 8. 9, 40 PGS. III. 8. etc.

35. *HGS*. I. 5. 16.

35a. Taittiriya Samhita, IV. 5. 10, 11.

36. A.B. Keith, *Religon and Philosophy of the Vedas*, vol. 31, p. 146, f.n. 5; Bhandarkar, Vaisnavism Saivism and Minor Religious Systems, pp. 106 ff.
37. *MBh*. Drona, P., ch. 18.
38. *MBh*., Santi, P., ch. 282.
39. *MBh*. Anusasana P., ch. 18.
40 *MBh*. Santi P., ch. 287.
41. *MBh*. Anus. P., ch. 17.
42. *MBh*. Drona P., ch. 201.
43. *MBh*. Anus. P., ch. 17.
44. *MBh*. Santi P., ch. 284, cf. Saup. P. ch. 6.
45. *MBh*. Santi, P., ch. 282
46. *Ibid*.
47. *MBh*. Santi P., chs. 14, 17.
48. *MBh*. Anus. P., ch. 17, Santi P., ch. 283.
49. *MBh*. Anus. P., ch. 17 Drona P., ch. 201.
50. *MBh*. Anus, P., ch. 14.
51. *MBh* Anus P., ch. 17.
52 *MBh*. Drona P., ch. 201
53. *MBh*. Santi P., ch. 284.
54. *MBh*. Anus. P., ch. 14.
55. *Ibid*.
56. *MBh*. Op.cit., Anus P., ch. 17.
57. Mbh. Ibid.
58. *MBh*. Santi P., ch. 284.
59. *MBh*. Drona P., ch. 201.

# 6

# Plants Health and Hygiene

*Svasthya*[1] is the Sanskrit term to signify the full meaning of the English word 'health'. In the Pali phraseology health is the *Sotthibhava* or well-being of the self. The well-being of the self primarily consists in *drogya* or diseaseless condition of both the body and the mind. So one need not be surprised if corresponding to the Sanskrit *arogyasala*[2] (hospital, lit., the home for the cure of diseases) Sanskrit *drogyasala*[2] (hospital, lit., the home for the cure of diseases) we come across the Pali term *sothisala*(svasthi-sala).[3] But the word *drogya* itself is frequently met with throughout Pali literature, and it generally means *arogabhava,*[4] the diseaseless condition. *Arogya* is not, however, a mere negative term denoting merely the condition of the self free from diseases, for it also denotes a fit, or healthy condition. According to the Charaka Samhita *drogya* is that condition of the self free from diseases, for it also denotes a fit, or healthy condition. According to the Charaka Samhita *drogya* is that condition of the self which is characterized by harmony among the elements *(dhatusamya,*[5] the elements constituing the entire self. In the language of the Susruta Samhita roga or vyadhi, the opposite of arogya, is a condition of the self which is attended with afflictions (*duksa-samyoga*).[4] Accordingly *arogya* is that condition of the self which is attended with the feeling of ease *(sukha-samyoga)*[7] Combining all these, roga may be defined as that condition of the self (constituted of body, mind, and the rest) when condition of the self (constituted of body, mind, and the rest) when all the parts and organs do not function together in harmony, and when it is attended with the feeling of unease, and arogya is that condition of the self when all the parts and organs function together in harmony and is attended with a feeling of ease.

That this was the accepted definition of disease and health is evident from Buddha's definition of *dukkha and sukha,* which are in some sense or other nothing but another set of terms for *roga* and *arogya.* According to Buddha, *dukkha* is that condition of the self when all the elements do not work together in harmony, in concord *(dhatu-vissandanam dukkham),* and *sukha* is that condition f the self when all the elements work together in harmony *(dhatu samsandanam sukham).* It goes without saying that in the Buddhist terminology *sukha* carries with it the sense of ease (sata), both bodily and mental (*kayikam cetasikam*), and the term dukkha carries with it the sense of unease (asata).[8] According to Charaka as there are kinds or degrees of discord so thee are kinds or degrees of harmony.[9]

In the Pali Suttas one meets with as many as five different words to comprehend the general idea of personal well-being, and all of them occur in the usually phrase: *appabadham appatankam lahutthanam balam phasuviharam.*[10] The five words are: (1) *appabadha*[11]—little or no illness, (2) *appatanka*[12]—little or no fear, or good health, (3) *lahutthana*[13]— lightness of body, (4) *bala*[14]—strength and (5) *phasuvihara*[16]—comfort, ease. The general idea of personal well-being was sought to be comprehended also by these *four* terms: (1) ayu-longevity, (2) vanna-brightness, (3) sukha-ease and (4) bala-strength.[17] According to the Dhammapada Commentary, of the four—ayu, vanna, sukha and bala, the main term is ayu, main because it is on the possibility of ayu that the possibility of the remaining three ultimately depends: *ayunhi vaddhamane yattakam kalam tam vaddhati tattakam itarepi vaddhanti yeva.* And ayu, according to the Charaka Samhita, means the organic interconnection of the body,the senses, the mind and the soul.[18] Thus the maintenance of health is the maintenance of life itself in different degrees of vigour.[19] The comprehensiveness of the meaning of *arogya* my be realized if one interprets it in the light of the then prevalent scheme of thought which is met within the Pali Canon in respect of evil and good *(akusala and kusala.*[20] The cure of present ill, the prevention of the future occurrence of ill. The maintenance of the present weal and the future growth of weal. Going by this scheme of thought any Science of medicine or any System of treatment claiming to be comprehensive has got not only to furnish prescriptions for the cure of the present disease, but also for the prevention of the recurrence of diseases, the maintenance and enhancement of the present health. In order to attain *arogya* in this comprehensive sense, two things are necessary: (1) the choice and application of drugs, and (2) the choice and application of the principles of hygiene. Medicines are primarily intended for curing diseases and remotely for increasing the strength and vitality.[21] and hygiene to prevent the recurrence or the emergence of fresh troubles.[22] The term hygiene is mainly concerned with ahara and vihara,[23] the rules of diet and the rules of personal behaviour conducive to health.The term vihara connotes the whole of personal behaviour in respect of personal cleanliness, residence, movements, exercises, dressing and the rest. All these rules of health are sought to be comprehended in Ayurveda by the term *svathavrtti.*[24]

Here, however, we are to deal only with that narrow aspect of health and hygiene in respect of which man has to depend plants and plant-products. We propose to deal with the subject under the following heads: (1) Medicine in General, (2) Personal Hygiene, (3) Some Hygienic Measures in General and dental Hygiene in Particular, and (4) Hospital and its Requisites.

### Medicine in General

The subject of development of the Science of Medicine does not concern us here. It may suffice for us just to point out that beginnings of this science may be clearly traced in the Vedic hymns, especially in the hymns of the Athara-veda, and that it reached a high state of perfection in t he treatises of Charaka, and Susruta as well as in the Astangahrdaya. The Indian word *ausadha,* denoting medicine, is derived from *osadhi,* meaning perennial herb. The other term *bhesaja* primarily denotes vegetable drugs. The latter term, however, came in course of time to have a much wider extension to denote all kinds of medicine, whether derived from plants, or from animals or from minerals or from minerals (Charaka Sam. I. i. 35: p. 47). The word *mula-bhesajja,* as used in the milinda Panha (p. 43) means 'regimen' or substances that are likely to increase vitality, the five such substances being *sappi, navanita, tela, madhu and phanita.*

With the Vedic Sages the plants were the main, if not the only source of medicine as will appear from the following quotations:—

"Mother (of mankind), a hundred are your applications, a thousand-fold is your growth, dó you who fulfill a hundred functions make this my (people) free from disease."[25]

"the universal all-pervading plants assail (disease) as a thief (attacks) a cowshed; they drive out whatever infirmity of body there may be." 10

"The plants falling from heaving (divine origin) said: The man whom living e pervade, will not perish."17[26]

The same mental preoccupation about plants in connection with Healing Science is patent also in Asoka's Rock Edict II, in which the Buddhist Emperor of India in speaking of the arrangements made by him for two kinds of treatment, thought only of plants and herbs, fruits and roots, i.e., substances derived from the vegetable kingdom. Coming to the still later period, we see that the Charaka Samhita palpably mentions no fewer than 36 kinds of medicine prepared from plants and plant-products, such as, roots, bark, wood, exudation, stem, sap, ash, latex, fruits, flowers, oil (essential and fixed), thorns, emergences, hairs, leaves, bulbs, corms, buds and the rest.[27]

An outstanding feature of the Hindu System of Medicine was the application of fresh drugs as opposed to synthetic drugs used in modern times. That the supremacy of this method of administering fresh drug is amply verified by what Dr. Ischirch is reported to have said: "We may assuredly hope that medicine, when it has thoroughly ruined digestion with synthetic drugs used in modern times. That the supremacy of this method of administering fresh drug is amply verified by what Dr. Ischirch is reported to have said: "We may assuredly hope that medicine, when it has thoroughly runed digestion with synthetical medicines, and tested all the organs of the animal body, will return to the most an ancient remedies of mankind, the medicinal plants and drug for the utility of which the experience of the thousand years vouches."[28]

## RULES OF PERSONAL HYGIENE[29]

With regard to personal hygiene the Susruta Samhita lays down certain daily rules of conduct, opinion that an intelligent man desiring perfect health and a sound body sho carefully observe them:

Acting upon these rules one is to leave one's bed early in the morning, brush one's teeth and scrape one's tongue[30] before washing one's mouth and eyes with a decoction of the bark of trees with milky juice, or with milk or Billodaka, or amalaka water, or with cold water alone. In the opinion of Susruta the face becomes handsome, and the eyesight stronger by such washings. After washing the eyes in the prescribed manner, one is to apply anjana collyrium) which has the virtue of protecting the eyes from diseases:

(Before going out for work) one is to chew betel-leaf alongwith other ingredients, this having the virtue of cleansing the mouth imparting a sweet aroma to it, enhancing its beauty, strengthening the voice, clearing the tongue, the teeth, etc.:

At the time of bathing one is to anoint first his head with oil '(siro ) in order to prevent affectations of the head. The oil is a means of making the hair grow luxuriantly and imparting thickness, softness and a dark gloss to it. After this one is to anion his body ( ) with oil,[30] this being recommended as a means of imparting a glossy softness to the skin, and giving a tone to the system. After this one is to take one's bath ( ) in cold water, this having the efficacy of removing somnolence, inordinate bodily heat and fatigue, allaying thirst, stopping itching ad excessive perspiration, producing a fresh relish for food, removing all bodily impurities, clearing the sene organs, gladdening the mind, purifying the blood, increasing the appetite, destroying drowsiness, and increasing semen. As for the difference between the effect of applying cold water and that of applying hot water on the head, the Susruta says that cold water so applied invigorates the eye sight, while hot water impairs it:

Immediately after bathing one is to comb one's hair in order to improve its growth, remove dandruff and dirt and destroy parasites of the scalp. Alongwith combing one's hair is to proceed dressing which also includes *anulepana, alepa and prasa dhana.* The anulepana, i.e., anointing the body with scented paste, is to be restored to for removing fatigue, foetor and perspiration and giving the body a lovely appearance. The alepa, i.e., besmearing the face with scented paste or powder, is recommended as a means of imparting steadiness to the eyes, producing a broad and graceful contour of the cheeks and mouth, preventing their disfigurement, etc. And the prasadhana, i.e., putting on of gems, flowers, clean clothes, etc., is prescribed as means of enhancing the beauty of the person, and maintaining cheerfulness of mind:

One is to take one's food[31] that is wholesome. By wholesome is meant that kind of food which can nourish the body, gladden the hear, invigorate the system, maintain the bodily strength, increase appetite and vitality, improve the memory and increase the energy and span of life:

One is to take daily exercise that make he body stout and strong, help the symmetrical growth of the limbs and muscles, improve the complexion and digestive powers, prevent laziness and render the body light and bright, firm and compact:

The rules of personal hygiene in Ayurveda also include:—

(a) Affusion of the body ( sarvodraparimok)—which removes fatigue, sets up fractured bones, alleviates paindue to burns, scalds, bruises and lacerations, and subdues the actions of the deranged vayu:

(b) Massage—which is recommended as a means of reducing the fat, smoothing and cleaning the skin and imparting a firmness to the latter. Megathenes noted Indians taking to this practice by passing rollers of ebony over the skin (Mc. Crindle, p. 69).

(c) Friction ( ) and rubbing ( ) of the body with medicinal powders: *Udgharsana* is recommended as a means of curing itches, rash and eruptions; and *utsadana* specifically for the improvement of the complexion of the females

(d) Washing of the feet—This should always be done with cold water, and as a result semen increases, heart is gladdened, fatigue disappears, vision becomes clear and the soles (of the feet) free from dirt and local diseases:

(e) Application of oil to the feet — which brings on sleep, invigorates sight, softens the skin, removes fatigue, etc.:

(f) Use of foot-wear — which is recommended as a means of protection against diseases, and making walking easy and pleasant. Walking without shoes is definitely perilous to life and health, and is always attended wit the danger of impaired vision:

(g) Shaving of hari and pairing of nails— these when periodically done will make a man cheerful, increase his energy and impart a lightness to his body:

(h) Wearing of turban— This is recommended as a protection against wind, dust, sun and strong light; it also helps the luxurious growth of hair, etc.:

(i) Use of umbrella—This should be used as a protection against rain, wind, dust, dew and sun. It also improves one's energy, eyesight and complexion:

(j) Use of stick— Which dispels the rear of dogs, snakes, tigers and like animals, and horned animals; increases one's energy, strength and patience; makes the mind firm and blood; acts as a proper support for the body and makes one fearless:

(k) Use of good bed and furniture ( ) — The Susruta recommends the use of a soft bed or seart while taking rest. This will remove fatigue, soothe the bodily vayu, bring on sleep and restore lost memory

(l) Fanning ( ) —Which is refreshing, keeps off flies and mosquitoes, arrests perspiration, removes fatigue, and fainting fits and alleviates the burning, scorching and parched, sensations.

(m) Gentle walk or stroll ( ) – Which 'ends to improve one's memory, strength, digestive capacity, and increases the duration of life:

(n) Putting on armour ( ) – Which improves one's strength, energy and complexion and gives a lustre to the body:

(o) Good sleep ( ) at the proper time – Which should always be done to improve the growth, strength, vigour and complexion of the body:

Some Hygiene Measures in General and Dental Hygiene in Particular

By hygienic measures in general we mean the measures recommended for the disinfection of rooms, clothes and the rest, as well as for the purification of water. Here the dental Hygiene is also accorded a special treatment for the simple reason that in almost all the treatises dealing with the subject of personal hygiene a special stress has been laid on this subject.

Disinfection: Disinfection in general, according to Charaka used to be effected by fumigating the clothes, or rooms and other places with the fumes of peacock's tail, bones of herons, mustard and ghee, According to Susruta, however, disinfection is caused by the burning of the following substances:Leaves of *nimba vrhati,* pepper, assafoetida, *jatamamsi,* seeds of cotton, etc., along with the tail of peacock–

and so on.

Sarngadhara gives the following formula of a compound by the use of which the pests of a room can be destroyed; If a room be fumigated with a compound made up of the flowers and fruits of *arjuna* tree, *laksa, haricandana, guggula,* roots of white variety of *aparajita, vidanga,* mustard and incense, serpents, rats, bugs and other insectiferous pests and flies are destroyed.

Purification of polluted water and drinks; Purification of polluted water, according to Susruta, could be done in seven different modes, such also, by the immersion in it of *kataka* fruits, *gomeda* (a kind of gem), the roots of lotus plants, algae, mosses and other water weeds, of a pearl or a crystal, or by straining through a piece of linen.[2] The water of a tank or well was to be purified by a compound of anjana and like things thrown into its water. This compound when thrown into a tank, the water of which is dirty, pungent, or otherwise disgustful to the taste, or of bad smell, would make the same water transparent, sweet, fragrant and invested with other good qualities.[3]

Bhavamisra, a later day authority, gives the following direction for the purification of pulluted water – ( ):

Purified water is then scented with the flowers of *nagakesara, campaka, utpala, patala* and others– Susruta sam. I. xiv, 12; p. 162.

The brhatsamhita gives the following recipe for scenting water – cardamom, *tejapatra,* and *rasanjana,* honey,pepper *nagakesara* and *kustha*— all these substances are to be taken in equal shares and kept under water, and after a few days the water will be nicely scented. For this purpose we now use extracts of *ketaki* or rose flowers.

*Dental Hygiene.*—The tooth-brush, the tongue-scraper, and the tooth-pick are the three main requisites of dental hygiene. according to the Susruta, the tooth-brush, an the tongue-scraper cleans the mouth, remove the bad smell and enable a man to eat with relish.

He then recommends the use of branches of such trees as have an astringent, or sweet, or bitter, or sour taste for tooth-brushes. Each group is again illustrated with one example, thus—*nimba,* is best amongst the bitter, *khadira* among the astringent, *madhuka* among the sweet and *karanja* amongst the sour.—

A later day authority, Bhavamisra (I.i), is more elaborate in the details. He says: *Arka* is conducive to strength, *vata* to brightness of appearance. *karanja* to victory, *plaksa* to wealth and prosperity, *vadara* to sweet speech, *khadira* to s weet scent of mouth, *vilva* to immense wealth, *udumbara* to truth of speech, mango to recovery, *kadamba* to intelligence and memory, *campaka* to steadiness, *sirisa* to name, fame, longevity, health and prosperity, *apamarga* to intelligence, memory, foresight, strength and voice, *dadima* to beautiful appearance, *kakuva* and *kutaja* to same as before, and *jati, tagara and mandara* dispel bad dreams:

According to the same authority, one should scrupulously avoid *guvaka, tala, hintala, ketaki, urhattrna kharjura and narikela:*

The Vinaya Cullavagga (v, 31, I; S.B.E. 20, 147) enumerates the following disadvantages arising out of not using tooth-sticks: (1) it is bad for eyes; (2) mouth becomes bad smelling; (3) the passage by which the flavours of the food pass become impure; (4) bile phlegm get into the food, and (5) food does not taste well.

The Brhatsamhita, which devotes a whole chapter to the subject of tooth-brush, says that there may be a thousand varieties of tooth-brush according t the nature of the plant whether it is a creeper, or a climber, or a shrub, or a tree. It then gives the following direction as to the selection or rejection of plants for tooth-brush: one should not make tooth-brush of trees the origin of which is not known, of trees that are with leaves,of which knots are equal, or those that are in flowers, or have their upper parts dry,or those that have no bark at all. One is blessed with divine prowess in his body if one cleanses his teeth with tooth-blessed with divine prowess in his body if one cleanses his teeth with tooth-brush made of the wood of such trees as *vaikantaka, sriphala and gambhari;* one gets a beautiful wife if one's teeth are cleansed by the wood of *ksema* tree, increase of wealth if with that of vata tree, increase of energy if with that of *arka* tree, of children if with that of *madhuka*tree and done becomes very popular if one cleanses his teeth with a brush of *arjuna* tree.

One is blessed wit the favour of the goddess of wealth if one prepares his tooth-brush out of the wood of *sirisa and karanja* trees, and one has all one's ends served and all desires fulfilled if one clearness his teeth with tooth-brushes of *plaksa* tree. Cleansing one's teeth with a brush of *jati* wood leads to the attainment of honour and that with *asvattha* wood leads to the attainment of paramountcy.

Preparation of a tooth-brush out of the wood of plum and *vadara*trees ensures recovery (from diseases) and longevity. a man's wealth is increased if his tooth-brush is made of *khadira and vilva* wood, and all his desired things are obtained if he makes use of a brush of *atimuktaka and kadamba* wood.

Cleansing one's teeth with a brush of *nimba* wood leads to attainment of wealth, that with one of *kravi* wood to the attainment of food and shelter, and that with one of *bhandari* wood to the attainment of food in excess, and that with one of *sann* and *arjuna* wood to the destruction of one's enemies; and that with one of *syama* plantalos to the annihilation of one's foes. Preparation of a tooth-brush out of the wood of *sala, asvakarna, devadaru and vasaka* ensures honour and that with the wood of *priyangu, apamarg, jambu,and dadima* guarantees popularity. 2-7.

We find the following recipe for scenting a tooth-brush: Four parts of jatiphala, two of talisapatra, one of cardamom, and three parts of camphor—all these are to be compounded, and with the compound the toothbrush mentioned above, should be besmeared and dried in the sun. "If one cleanses one's teeth with this tooth-brush his face becomes beautiful and spotless and fragrant. and one' words become charming."

In the Daily Practices of the Hindus (S.B.H. XVI) the following plants are recommended for the selection of tooth-brush: Datepalm, kadamba, karanca, banyan, tamarind, bamboo, mango, nimba, vilva, akanda, and fig trees.

After cleansing the teeth and gum with tooth-brush: Detepalm, Kadamba, karanca, banyan, tamarind, bamboo, mango, mimba, vilva, akanda, and fig trees.

After cleaning the teeth and gum with tooth-brush, which is generally a small branch of about 6-9 inches long, it is split open and a half of it is used as a tongue-scraper, and then the whole thing is thrown away.

Regarding the use of a tooth-pick after meals, the Susruta lays down: particles of food adhering to the teeth or lodged between the teeth should be extracted by means of a tooth-pick, which is ordinarily a stick of grass, there wise, these (food particles) will decompose and the mouth would be smelling very badly:—

All these elaborate injunctions point to one fact namely,that the ancient Indian used to take particular care of his or her teeth and suffered less from diseases. For this he no to only used tooth-brush in the early morning, but also every time after his principal meals. The practice was so much in vogue that foreign observers could not afford to overlook this daily practice in India. Yawn Chwang, for instance, notes: After eating they cleanse their teeth with a willow stick, and wash their hands and mouth.[1] But I-Tsing gives a more detailed account of this practice under the title "Cleansing after meals." Says he The Indians clew tooth-wood in the mouth, let the tongue as well as the teeth be carefully cleansed and purified after meals. They chew tooth-woods and clean the teeth with them and rub off the dirt of t ehtongue as carefully as possible. Tooth-wood is made about twelve finger-breadth in length, and the minimum length being eight finger-breadth. Its size is like the little finger. One of its ends is chewn softly and the teeth), and bending it (tongue-scraper) the tongue is rubbed. As tooth-picks small sticks of bamboo or wood flat as the surface of the little finger, and sharpened at one end may be used.

About the source, he says: Tooth-brushes are made out of a large piece of wood, small stem of tree, branch of an elm, and a willow tree (see Jataka 1, 80; Mahavamsa, 23) or any sweet smelling wood, or root, or a creeper. The best one is bitter, astringent or pungent in taste, or one which becomes like cotton when chewed. The rough root of the northern Burr-weed is the most excellent. It hardens the teeth, scents the mouth, helps to digest the food, or relives heart burning. If this kind of tooth cleaner be used the bad smell of the mouth will go off after a fortnight. A disease of canine tooth,or toothache will be cured after a month. Be careful to chew fully and polish the teeth cleanly. Toothache is very rare in Indian owing to their chewing the tooth-wood.Even infants of three years' old are taught how to do it.[2]

**Hospital and its Requisites**

Corresponding to the Sanskritt *arogysala* (a home for the cure of diseases) we have in Pali *gilanasala* ( a hall for the sick). The term *gilanasala* is met with in the earliest of the Pali texts and it also occurs throughout the later Pali literature.[3] Besides *gilanasala,* we come across *sutthisala* as an exceptional term to denote the Indian idea of a hospital. Some of the kings of Ceylon are credited in the chronicles of Ceylong (Mahavamsa) with the foundation of both hospitals and maternity homes.

The real impetus to the building of hospitals and similar institutions was given by king asoka who in his Rock Edict II claims to have made arrangements for two kinds of treatment (*dve cikichad kata*) one suitable for men and the other for animals. If we can take him at his own words,he made those arrangements not only throughout his wide dominions but throughout the principalities of his allies as well the territories of the five Greek rulers in the north-west, and the kingdom of Chola, Pandya, Satya putra, Keralaputra, in the south as far as Tamraparni.

The Arthasastra (Bk. II, ch. iv, 55) expressly recommends the construction of *bhaisajyagrha* (another term for hospital) at the north-west corner of the compound of a fort. The same Arthasastra refers even to an organization similar to if not the same as, the modern Red Cross Society. The passage in question runs thus: That is, "physicians with surgical instruments, machines, remedial oils and bandages in their hands: and women (nurses) *drogyasala* is praised in the Vaidyaka sastra as a monumental work of piety.the hospital was to be equipped with the following requisites: (1) a good stock of important drugs, (2) expert physicians, (3) and a good store of food and regimen. The following are mentioned as qualities of expert physicians: that they will be well versed in the science, wise and adept in diagnosis and choice and application of drugs, as well as in the prescription of proper diet.[2] According to the susruta the surgical ward is to be equipped with (1) Medicine, (2) necessary articles of diet, (3) dissecting apparatus (sastras and anusastras), including splints, (4) aspirator to drain off blood, tc., and (5) suture material and surgical box.

### 1. *Medicine*

The first and the most essential requisite of the hospital is medicine. According to the Vaidyaka Sastra, the *drogyasala* was to be equipped with all-important medicines—It will be preposterous to suppose that the stock of medicine consisted only of drugs derived from plants and plant-products. In the later developed stage of the system of treatment the drugs were prepared from the vegetable, animal, as well as mineral sources. But whatever the main ingredient of an Indian drug, in actual application it was generally accompanied by a fresh drug ( ) for which the primary source is plant.[1]

### 2. *Articles of Diet*

The next requisite of a hospital, according to the Vaidyaka Sastra, is a sufficient store of wholesome food and regimen ( ). So far as plant products are concerned, the articles of diet consisted of different kinds of rice, such as *sali and sasthika, mudga, masa, yava, tila, kulattha, vadara, mrdvika, parusaka, abhaya, amalaka, vibhitaka* and other classes of medicaments. (Charaka samhita (I, XV). As far as the pure medical side goes the Charaka Samhita (I, xv) gives an elaborate description of a nursing homegiving a detailed direction as to the construction of the home and its equipment as regards the personnel, the drugs, provisions of foodstuff, sanitation, and the method of treatment of diseases and nursing the patients, both during the attack and convalescence.

### 3. *Surgery*[2]

That the science of surgery was far advanced in Susruta's time is definitely evident from a clear instruction given for the dissection of the dead body in the Susruta Samhita: "Any one desirous of acquiring a through knowledge of anatomy should prepare a dead body and carefully observe (by dissecting it), and examine its different parts. For a through knowledge can only be acquired by direct personal observation." The picture of the practice seems almost modern. Here it is: "the body must first be disemboweled, and wrapped round with the outer sheaths of *munja* grass, or hemp, kusa or other grasses, then kept immersed in still water for seven days, after which the medical student should proceed to remove the layers of the skin with a carefully prepared brush made of the grass roots, hairs, kusa blades, or strips of split bamboo, and carefully observe with his own eyes all the various different organs, external and internal, beginning with the skin as described before:. But the above was the final course in practical survey. Before that the students were asked to try their knives repeatedly on artificial objects resembling the diseased parts of the body before undertaken an actual operation. Incision, for example, was practiced on *puspaphala, alabu,* water-melong, cucumber of *ervaruka or trapusa;* enesection was practiced on the stalks of water lilies; the art of probing and stuffing on worm-eaten wood, bamboo reeds, or on the mouth of dried *alabu*; extraction of solid bodies by withdrawing seeds from the kernel of vimbi vilva or panasa, scraping on wax spread on a salmali plank and suturing on pieces of cloth, skin or hide. There were 4 methods by which dissection or operation could be done namely, with apparatus, instruments, alkalies and fire

(a) Surgical apparatus (*yantra*)—Alabuy (gourd) yantra[1] is mentioned by both the Charaka (VI, vii) and the Susruta (I, xiii)— Its mode of preparation is described. It used to be employed for draining blood or phlegm from the body like modern cupping glosses.

(b) Surgical instruments (*Sastra*)— Bamboo-skin was one of the materials out of which these instruments used to be made Some of these instruments were named and fashioned after some plant parts, such as, *masura-dala mukha, jambhavavadana, mukulagra, malatipuspa-vrntagra—salaka* (rod-shaped) yantras. Some knives were even named as *utpalapatra, kusapatra, vrihimukha,* and *vetasapatraka* (S.S.I, vii).

(c) Substitutes for cutting-instruments *(anusastras)*— These were subsidiary apparatus made out of plants and other things used for dissecting purposes. These were

Of all these Anusastras ksara (caustics) are the best etc (S.S.I, ch. xi, 2; Charaka Sam., Cikitsasthana, ch. xxvii, 26, ksara-agada; Roy's Hist. Hindu Chemistry I (1902), p. 17). For the preparation of alkalies, their uses, different strengths, eternal and internal use the Susruta (I, xi) and Rasarnava ( ) vii, 12-13), may be consulted. The sharp action of alkalies may be relieved when mixed with acid.—

(d) Splints—were also in use for surgical purposes, as they are now. Susruta prescribes the bark of the following plants for the purpose:

Bhavamisra adds *nicula* to the above list. Vagbhata II, gives the following list with a direction as to the application of the splint, thus:

That the ancient splints made of bark were in some respects more efficient than their counterparts will be evident from the following quotations: "Dr. Jacobi of Dublin says that he has seen an excellent splint made from the fresh bark of a tree taken off while the sap is rising. It fits admirably, says he, just like a pste-board soaked in water (Hamilton's Fractures and Dislocations, 5the ed., pp. 50-51). Dr. C.C. Jewt recommends for he same purpose the bark of Leriodendron, or tulip tree (the 20th Mass. vol. I)"[2]

***Suture Material***

For this susruta recommends the following:

After properly suturing the ulcer 'should be covered over with cotton and dusted over with a pulverised compound consisting of th e powders of *priyangn, anjana, yastimadhu and rodhra,* or with the ashes of a burnt, piece of *ksauma* cloth, or with the powders of the *sallaki* fruit.' Finally the bandage is completed by wrapping with *ksauma* cloth, patrorna, antarvalkala, alabu sakala, lata-vidala, and tulaphalam among various other articles.

**5. *Surgical Box***

The Susruta finally recommends that proper care should always be taken of the surgicala instruments, and to preserved the sharp edges of the cutting instruments heath, or receptacle made of the wood of salmali should always be used:

S.S.I., vii.

# 7

# Agriculture and Rural Development in India

From a nation dependent on food imports to feed its population, India today is not only self-sufficient in grain production but also has a substantial reserve. The progress made by agriculture in the last four decades has been one of the biggest success stories of free India. Agriculture and allied activities constitute the single largest contributor to the Gross Domestic Product, almost 33 percent of it. Agriculture is the means of livelihood of about two-thirds of the workforce in the country.

This increase in agricultural production has been brought about by bringing additional area under cultivation, extension of irrigation facilities, the use of improved high-yielding variety of seeds, better techniques evolved through agricultural research, water management, and plant protection through judicious use of fertilizers, pesticides and cropping practices.

## Crops

The 1970s saw a multi-fold increase in wheat production that heralded the Green Revolution. In the next decade rice production rose significantly; in 1995-96, rice production was 79.6 million tonnes. The total grain production crossed 211 million tonnes in 2001-02, a big leap from 51 million tonnes in 1950-51.

## Irrigation

To carry improved technologies to farmers, a National Pulse Development Programme, covering 13 states, was launched in 1986. Efforts to boost pulse production were augmented further by the Special Food Production Programme. In 2001-02, pulse production was 13.52 million tonnes. With some states offering more than the statutory minimum price, sugar cane production also received a boost, and in 2001-02 a record 292.2 million tonnes was registered.

As efforts continued to increase the irrigation potential in the country, the last 40 years saw the gross irrigated area reach 85 million hectares. Flood forecasting has become an important activity over the years. Over 500 hydrological stations collect and transmit data through 400 wireless stations for issuing forecasts for 157 sites. About 5,000 forecasts are issued in a year with 94 percent accuracy. The country also receives international support, with the World Bank as a primary source, for developing its water resources. International cooperation is also envisaged in setting up a National Centre for Information on Water and Power. As there is a broad seismic belt in the country, particularly along the Himalayan, and the Kutch region and parts of Maharashtra, a scheme is being evolved to collect all data on seismic activity at various dam sites.

## Fertilizers

The fertilizer industry in India has grown tremendously in the last 30 years. The Government is keen to see that fertilizer reaches the farmers in the remote and hilly areas. It has been decided to decontrol the prices, distribution and movement of phosphatic and potassic fertilizers. Steps have been taken to ensure an increase in the supply of non-chemical fertilizers at reasonable prices. There are 66 fertilizer quality control laboratories in the country. Since bio-fertilizers are regarded as an effective, cheap and renewable supplement to chemical fertilizers, the Government is implementing a National Project on Development and Use of Bio-fertilizers. Under this scheme, one national and six regional centers for organizing training, demonstrating programmes and quality testing of bio-fertilizers has been taken up.

It was a challenging decision of the Government to take Bombay High gas through a 1,700 km pipeline to feed fertilizer plants located in the consumption centers of North India. However, the major policy which has ensured the growth of the fertilizer industry is the thrust on accelerating fertilizer consumption by fixing, on the one hand, low and uniform price for fertilizers, and on the other hand providing the manufacturers adequate compensation through the retention price and subsidy scheme. As expected, fertilizer nutrient demand has gone up from 0.29 million tonnes in 1960-61 to 16.7 million tonnes at the end of 2000-01, compared to 12.15 million tonnes during 1992-93.

## Fisheries

Fish production achieved an all-time high of 5.6 million tonnes at the end of 2001-02. Programmes that have helped boost production include the National Programme of Developing Fish Seeds, Fish Farmers' Development Agencies and Brackish Water Fish Farmers' Development Agencies. The Central Institute of Fisheries Nautical and Engineering Training trains the necessary manpower.To diversify fishing methods and introduce processed fish products on a semi-commercial scale, an Integrated Fisheries

Project has been launched. A National Fisheries Advisory Board has also been established.

## Food Processing

The Ministry of Food Processing Industries, set up in July 1988, is the central agency of the Government responsible for developing a strong and vibrant food-processing sector with a view to create increased job opportunities in rural areas, enable the farmers to reap benefit from modern technology, create surplus for exports and stimulating demand for processed food.

A new seeds policy has been adopted to provide access to high-quality seeds and plant material for vegetables, fruit, flowers, oilseeds and pulses, without in any way compromising quarantine conditions. Initiatives have been taken to encourage private sector investment in the food processing industry.

## Agricultural Research

The apex body for education, research, training and transfer of technology in the field of agriculture is the Indian Council of Agricultural Research (ICAR), established in 1929. India's transformation from a food deficit to a food surplus country is largely due to ICAR's smooth and rapid transfer of farm technology from the laboratory to the land.

ICAR discharges its responsibilities through 43 research institutes, four national research bureau, 20 national research centers, nine project directorates, 70 all-India coordinated research projects, and 109 Krishi Vigyan Kendras (farm science centers).

Besides, the programme of Agricultural Education is coordinated by ICAR with the curricula and other normative guidance given to the 26 agricultural universities and four national research institutes.

## Oilseeds Production

A Technology Mission on Oilseeds was launched in 1986 to increase production of oilseeds in the country and attain self-sufficiency. Pulses were brought under the Technology Mission in 1990. After the setting up of the Technology Mission, there has been consistent improvement in the production of oilseeds. The oilseed production, which was 108.3 lakh tonnes in 1985-86, has increased to 247.30 lakh tonnes in 1998-99, the highest production so far. However, due to unfavorable weather conditions prevailing in the major oilseeds-growing states, the oilseed production during 1999-2000 and 2000-01 suffered some decline.

The increase in production was largely contributed by soybean, rapeseed and mustard. Production of pulses has seen many ups and downs, which is expected to be checked under the mission. The country grows mainly nine oilseeds, with groundnut, rapeseed and mustard accounting for 62 percent of total production. Lately, soybean and sunflower have shown major growth potential.

## Drinking Water

A Technology Mission on Drinking Water and Related Water Management has been constituted to cover the residual problem villages and provide potable water at 40 liters per capita per day, and 70 liters per capita per day in desert areas inclusive of 30 liters for cattle. The mission is tackling the problem through 55 mini-missions in project districts and countrywide problem-oriented sub-missions. A Village Level Operation and Maintenance (VLOM) pump called India Mark-II has been developed and is being exported to 40 countries.

## The Nature of Agricultural Development in India

That the post-independence period marks a turning point in the history of Indian agriculture is clear from the fact that compared with a rate of growth of less than 0.5 per cent per annum during 1904-5 to 1944-45 the agricultural sector recorded an annual growth rate of 2.7 per cent during 1949-50 to 1983-84. This growth has been achieved as a result of high priority accorded to agriculture. The policy makers adopted a two fold strategy for regenerating agriculture immediately after independence. The first element of this strategy was to implement land reforms in order to remove institutional bottlenecks and the second element was to undertake massive investment in irrigation and other infrastructure in order to update the existing agricultural technology.

The most important event in the social history of India was land reforms enacted and implemented during the mid-fifties. Simultaneously, technological upgradation has also played a crucial role in augmenting agricultural production. It is rather curious that whereas land reforms have been studied in detail, the role of technology and the relations between technology and institutions have, by and large, been ignored by scholars in this country. Whereas land reforms have determined the pattern of land distribution and thereby the distribution of incomes and land relations, technology has determined the production potential within the context of the given institutional framework. Since technology does not exist in a vacuum but operates within a given institutional and social context, it is important to study their interaction. It is true that institutional changes by themselves can also lead to increased output by releasing forces of production. For example, the abolition of intermediaries in India did lead to substantial increases in output during the mid-fifties. But without technological innovations, the potential for higher growth is limited and soon gets exhausted. To maintain a tempo of high growth, huge investments in new technology are essential. It needs to be underlined that without investment in irrigation, roads, electrification and other infrastructure, research and development and extension services, large increments cannot be achieved in agricultural output. It is, therefore, no wonder that the fastest growth has not taken place in states like Kerala and West Bengal where land reforms have been quite radical but in Punjab and Haryana which are not known for any radical land redistribution. The point to be stressed is that whereas investment in new technology determines the frontiers of growth, the nature of growth and sharing of gains of new technology are determined by the institutional structure, in particular the pattern of land ownership. The latter, in turn is a direct consequence of changes brought about through land reforms. It needs to be underlined that technology and institutions are inextricably linked and cannot be easily separated. It appears that a great deal of confusion has been created by scholars who have classified the Indian agricultural development into two periods 1950-51 and 1966-67 and that between 1967-68 and 1983-84 and have characterised these as the land reform strategy adopted during the mid-fifties and the green revolution strategy adopted since the mid-sixties. It is further stated that whereas the first strategy led to equitable distribution, the latter has resulted in accentuating inter personal and inter regional inequalities. Obviously, this characterisation is unscientific because a clear cut dichotomy is sought to be created between technology and institutions. It is true that land reforms brought about important changes in most of the states in India during the mid-fifties.

The most important contribution of land reforms was abolition of intermediaries and giving land titles to the actual cultivators. This released productive forces and the owner cultivators put in their best to augment production on their holdings. Whereas land reforms were important in increasing agricultural production during the first period, it also needs to be emphasised that during this period major output increases also took place because of large investments in irrigation infra-structure. In short, one could characterise the first period as having irrigation technology. Thus along with institutional changes, irrigation technology was an important instrument of growth. Further, it is absolutely incorrect to say that this period was characterised by equitable pattern of growth. The most important limitation of land reforms in various states was its failure in implementation of land ceilings enactments. Consequently land ownership distribution continued to be highly skewed. Because of this even during the first period, the inter-personal inequalities continued to be very large. There is evidence to suggest that the major gainers of irrigation technology were the upper middle, the rich and large farmers. It is no doubt true that the extent to which land reforms resulted in abolishing the inter-mediaries did release production forces by removing institutional barriers and thereby led to increase in output.

But since large inequalities in land ownership were legitimised, the pattern of growth was such where the main agents of growth were the upper middle and rich farmers. Another point to be noted is that although green revolution technology did make a qualitative difference in augmenting yields, in a sense, it represented a continuation of the earlier irrigation technology. Given assured irrigation much higher yields became possible because of genetic break through in seeds. Highly accelerated output when super-imposed on iniquitous institutional framework has had a far reaching impact on agrarian structure and land relations.

For a proper understanding of the changing agrarian structure, the various strategies adopted for augmenting growth should be evaluated bearing in mind the highly skewed distribution of land ownership and land holdings. Soon after independence India found that the domestic production of foodgrains was not adequate to meet the domestic demand. it was justifiably considered humiliating for a country of the size of India to be going around with a begging bowl. Hence increasing foodgrain output and achieving self sufficiency in foodgrains become a matter of high priority for the policy makers. The Grow More Food campaign, the Community Development Programme and the Intensive Area Development Programmes were all attempts at regenerating Indian agriculture that had stagnated during the British period. Having created an institutional structure, it was but obvious that solutions had to be found within that framework. In this context food output could only be increased through programmes like JADP, through investment in infrastructure in already irrigated areas and through increasing dependence on the main agents of growth namely the upper middle and rich farmers. It was, therefore, inevitable that the growth under such circumstances be concentrated in irrigated regions and in absolute terms a major proportion of incremental income should flow to the rich and very big farmers. Looked at in this context, to blame green revolution technology for distorted pattern of agricultural development is totally mis-conceived and illogical. It could actually be argued that without the introduction of new technology, India could not have recorded high growth in agriculture and could not have achieved near self-sufficiency in foodgrains:

Major reasons for unequal pattern of development, it may be repeated was the creation of skewed land structure as a result of half-hearted land reforms.

The nature of distortions that have taken place in the agrarian structure and their implications for our polity may briefly be analysed. The following seem to be the major characteristics of agrarian scene in India:

1. Inter-personal inequality

2. Inter-regional inequality

3. Poverty.

A great deal of empirical research has been undertaken on all the above aspects. Some of the conclusions are summarised below:

1. Coming first to interpersonal inequality, almost all studies bring out there exists very large inequalities in the distribution of both income and assets in rural India. Among the cultivator households although income inequalities are generally highly correlated with land ownership/land holding inequalities, the extent of income inequalities are found to be slightly less than that of land holding inequalities. This happens primarily because the smaller farmers are able to put in more family labour on their holdings and are also able to earn relatively higher income from non-agricultural and allied agricultural occupations. All categories of cultivators seem to have recorded substantial increases in their income as a result of adoption of new technology in areas where new technology has been introduced.

It is the middle and small farmers rather than the large farmers, who seem to have derived relatively higher benefits out of the initial thrust of green revolution, mainly because of their ability to augment yields by adopting labour intensive methods by utilising their family labour in an optimum manner. With the intensification of new technology, the capitalist farmers are also tending to become as efficient as the small owners through optimum utilisation of capital assets. The existing studies indicate that the statistical measures do not show unambiguously that increases in income inequality have occurred as a result of the adoption of new technology. In fact if one looks at Ginni coefficients or inter-class concentrations one finds a slight decline in inequality after the advent of green revolution.

This notwithstanding, the income inequalities are found to be quite large. I would like to submit that the statistical measures of inequality are quite misleading. For example to suggest that inequalities have not increased is a matter of little consolation and does not capture the real situation. For example, for marginal farmer household more than doubling of income means that the family can now have two square meals a day which probably it could not afford earlier. For a large and rich farming household only doubling of income means that it can not only construct a new house and get children educated, but also that it can now acquire a political clout and purchase patronage. I would submit that this is what has happened. The political clout of middle and rich farmers has increased tremendously whereas the living conditions of a smaller and marginal farmers have also improved slightly in the regions where new technology has been introduced. Most of the studies also indicate that in slow growing regions where new technology has not spread, the condition of marginal and small farmers as also of landless labour continues to be quite dismal and many of them continue to live below the poverty line.

The story about landless labour is more complex. In general in regions where new technology has been introduced it has led to increases in labour productivity. This, in turn, initially led to higher bargaining power of the landless labour and increased incomes and wages for them. However, this has gradually been counteracted by increased mechanisation on the one hand and large rural to rural migration of landless labour on the other. These two factors have tended to dampen the increases in wage rates in the prime green revolution regions. For example, the largest data indicate that the wage rates in

Punjab have recently tended to stagnate or even decline. The other factors that have influenced the wage rates is the extent of non-agricultural employment, and also the food for work programmes.

Regional inequalities are equally notable. For example it is the north western states of Punjab, Haryana and Uttar Pradesh which have recorded consistently high growth rates since mid-sixties. On the other hand the Eastern region comprising of Orissa, Bihar and West Bengal has had a dismal performance. Except for Andhra Pradesh the recent performance of the southern region has not been very satisfactory. Being primarily dependent on rains the central region has demonstrated a high degree of instability in its growth performance.

A disaggregated districtwise analysis gives very startling results. For example 60 per cent of the incremental output achieved in the country during 1962-65 to 1980-83 was accounted for by only 56 districts (having only a quarter of the total cultivated area). At the other extreme 151 districts which claimed nearly half the cultivated area accounted for only 20 per cent of incremental output. Another 25 districts (with 7 per cent of national area) actually made a negative contribution to total output. The 56 highly productive districts are mainly concentrated in the states of Punjab, Haryana and Uttar Pradesh, although quite a few of these are also located in Andhra Pradesh and Gujarat.

Coming now to rural poverty only a few points needs to be made. Firstly it is established that rural poverty is inversely correlated with agricultural growth. Secondly, notwithstanding the recent data, the extent of rural poverty is extremely high. Thirdly, although anti-poverty programmes have tended to somewhat alleviate poverty in some parts, the vast amount of corruption and leakages and the incapacity to create permanent assets have made these programmes exorbitantly expensive. Nevertheless these programmes continue to be attractive for political reasons.

To sum up, the agrarian scene is not too comforting. Very large inter personal inequalities have given rise to deep discontent resulting in serious conflicts in many parts of India — notably in Bihar and Andhra Pradesh. The regional discontent is equally disquieting. Assam, Gorkha Hills and Punjab are the clear manifestations. It appears that with all these conflicts we are sitting on a powder keg. The emergence in this scenario of rich peasant leaders with strong regional affiliations has made the scene even more murky.

The important fact to remember is that basically such peasant leadership is bound to be backward looking, non-secular and tyrannical towards the rural poor. Their increasing clout and elevation from the state to national politics is bound to disrupt the national consensus so assiduously built by the national bourgeoisie.

But it is the national bourgeoisie that is primarily responsible for the emergence of present situation — in so far as they failed to build an equitable agrarian structure throughradical land reforms during the 'fifties. Now it is too late in the day as the political power fortress of landed interests has become almost impregnable.

The problem could have become manageable had there occurred rapid industrialisation and significant shift of labour force from agriculture to non-agricultural occupations. After all classical capitalist development did result in large rural to urban migration in all the western countries.

But in India the pace of industrialisation has been relatively slow and halting. This is because industrialization and modernisation require large amounts of investment. Until recently, the rate of investment has been quite low. The recent increase in investment has been accompanied by sharp increase in capital output ratio. Further, industrialisation in developing countries needs huge investment in research and development for building up their own technologies. Besides capital investment, it requires large outlay on science education. Another important reason for slow industrial growth has been the deficiency of domestic rural demand which in turn has been constrained due to slow growth of agricultural incomes in many parts of the country. One would have expected little diversification of labour force with the type of growth rate in manufacturing in India. But the labour absorption has been lower than expected because of increasing capital intensity.

Thus the increasing capital intensity of consumer goods industry has hampered rapid growth in manufacturing employment, despite lip service paid to small scale industries. The recent policy of import liberalisation has on the one hand created serious problem for the domestic capital goods industry, it has simultaneously accelerated the process of capital intensification. it is no wonder therefore, that even with eight per cent annual growth in manufacturing output since 1980-81, there has been virtually no increase in manufacturing employment.

Whereas unemployment is a major problem one cannot forget about the problems posed by increasing inequalities not only within the agricultural sector but also between urban and rural India and between organised and unorganised sectors of the economy. In India for example, the average income of workers in the organised sector was four times that of the unorganised workers. Further, among the unorganised workers, the agricultural worker had one third the income earned by the non-agricultural workers. This level of inequality has created the rural-urban divide along with great deal of tensions.

It needs to be underlined that equitable distribution cannot be reconciled with the path of capitalist development. It requires a revolution to abolish the feudal inequalities of land ownership. Since these were not abolished, the capitalist development in India has resulted in further aggravating inequalities because of increasing value of land, property, other assets and because of multiplication of money acquisition opportunities. The extent of inequalities can be gauged from the fact that during 1975-76, whereas the 20 per cent richest households accounted for 48.4 per cent of total consumption, the poorest 20 per cent households accounted for only 7 per cent. Equally serious are the dilemmas posed by the type of urbanisation taking place in the country. whereas urbanisation is a necessary concomitant of industrialisation, the pattern of urbanisation taking place in India is full of distortions. this is because despite Government intervention the land for housing has by and large been appropriated by the colonisers, the merchants and traders and affluent professionals and the migrant labourers from rural areas are forced to live in burgeoning slums.

In short, the problems of excessive dependence on agriculture, slow process of agricultural and industrial growth and ensuing inequality, unemployment and poverty are much too serious and demand careful thinking by all the enlightened and progressive sections of people in the country.

**History**

International Fund for Agricultural Development (IFAD), is a specialized agency of the United Nations, and was established as an international financial institution in 1977 as one of the major outcomes of the 1974 World Food Conference. IFAD's mission is to enable poor rural people to overcome poverty. Working with rural poor people, governments, donors, non-governmental organizations and many other partners, IFAD focuses on country-specific solutions, which can involve increasing rural poor peoples' access to financial services, markets, technology, land and other natural resources. India is one of the founding members of IFAD and has a permanent seat on its Executive Board. India is not only a leading donor among the developing countries for IFAD but also has the largest country programme of IFAD in the world. Working in close partnership with the Government of India and other donors, IFAD has been funding projects for rural development, tribal development, women's empowerment, natural resources management and rural finance. Since 1979, IFAD has financed 21 programmes involving highly concessional loan assistance of approximately US$565 million. As on October 31,2007 IFAD has 7 ongoing projects involving loan assistance of approximately US$ 200 million.

IFAD's strategy in India centres on improving rural poor people's access to economic and social resources. In all operations, IFAD emphasizes the importance of strengthening people's capacities to establish and manage their own institutions. It supports Self-Help Groups (IFAD is a pioneer for the SHG movement India), community institutions and village development associations in tribal and non-tribal areas.

These and similar groups participate directly in designing development initiatives and become progressively responsible for programme and project resources and management. Empowering women and other disadvantaged groups is a strategic priority. IFAD has played an important role in developing replicable models through its projects and acted as a catalyst for far-reaching innovative change. In areas such as microfinance and women's empowerment, IFAD-funded operations have tested institutional and technical innovations that have been ultimately scaled up by the Government.

Strategic activities in India include:

Providing access to microfinance services, which are particularly effective in empowering women, Improving livelihood opportunities for tribal communities and in semi-arid tropical areas, where better water management and new technologies for agriculture help reduce poverty, Introducing development activities in the intensely populated and impoverished areas. Developing partnerships with non-governmental organizations (NGOs) and the private sector to reinforce community-based organizations Exploring possibilities for linkages with other donors and institutions Promoting policy change through project activities.

IFAD has developed close working relationships with other donors and agencies in India. It has a particularly strong partnership with the World Food Programme (WFP) involving joint funding and supervision of projects as well as collaboration on advocacy and other initiatives. IFAD has also cofinanced programmes and projects with the Department for International Development (United Kingdom) and the World Bank. The strategic framework of IFAD also recognizes that IFAD must expand its engagement beyond the immediate impact of IFAD's projects and programmes to influence the direction and content of national and international poverty reduction processes. Thus it emphasizes building complementary partnerships and broad alliances to maximize IFAD's contribution to the international community larger poverty reduction effort.

IFAD's interventions in India also strive to reorient the mindset of development agents at all levels in government, financial institutions, NGOs and community based organizations (CBOs). A conscious attempt is made to bring a more sensitive and responsive approach to the design and implementation of development interventions and to build the capacities of the poor through establishing or strengthening their institutions.

## Early Programmes

The first 5 projects approved by IFAD for India were for irrigation and command area development and all were cofinanced with the World Bank. These projects were: Bhima Command Area Development, Rajasthan Command Area Development and

Settlement, Sunderban Development, Madhya Pradesh Medium Irrigation, and Uttar Pradesh Public Tubewells. Since 1987 IFAD changed its strategy in India in view of the need to tackle the issue of food security at the household level. A series of projects targeting specific communities groups were developed. The first in this line was the Orissa Tribal Development Project, which identified the tribal population of Kashipur in Orissa as the primary target group and adopted broad-based interventions in natural resource management and rural infrastructure. This was followed by the Tamil Nadu Women's Development Project, which targeted women on an exclusive basis and introduced micro-finance as the principal instrument for empowering women. It was followed by the Andhra Pradesh Tribal Development Project which aimed at fostering household food security by increasing food production and raising the incomes of tribal households. These projects have since been completed.

# 8

# Organic Farming

If we were to playa word association game and someone said "sustainable agriculture," the reply would more than likely be "organic farming." This point raises some interesting questions. Are they associated practices? Are they equivalent? To answer this question, we must turn to working defini-tions.

A sustainable agriculture system is one in which the goal is permanence achieved through the utilization of renewable resources. The permanance sought is not static, but dynamic. The latter is necessary because some resources (e.g., population increases and cost increases for a diminishing supply of fuel energy) are not truly controllable. Hence, the need is for a dynamic permanence-one that permits the attainment of optimal produc-tion through the application of increasingly efficient techniques.

Basic elements of sustainable agriculture are the conservation of energy, soil, and water. Essential to the achievement of conservation are agricultural practices that are directed toward renewability and non pollution of resources. A renewed resource that is contaminated with pesticides and unusable or a polluted environment that reduces crop productivity is counterproductive.

Such concepts do not imply a return to primitive or nonindustrialized agriculture with the complete abolition of agricultural chemicals. More effi-cient methods undoubtedly will depend on increased technology, a tech-nology that is energy efficient and nonpolluting, or nearly so. Return to a chemical-free, less technological agriculture would be costly in terms of increased human labor and, most likely, would result in decreased agri-cultural productivity. Essentially, sustainable agriculturists will have to bal-ance resource conservation and environmental maintenance against mainte-nance of optimal productivity and minimal labor input.

Agricultural chemicals and mechanization would still have a place in sus-tainable agriculture, but not as before. Use of less chemical fertilizers would reduce energy demands and pollution; fertilizers would be supplemented with legumes, crop residues, green manures, and readily available, local supplies of organic wastes. Their utilization would help further to reduce costs and energy demands, remove a source of potential pollution, and, more importantly, renew soil organic matter. Practices to control erosion would be maximized to make soil a renewable, rather than a minable re-source.

Pesticides would still be used, but only as a component of integrated pest management systems that rely primarily upon effective natural controls. Pesticides would be used only when natural controls were not effective and selection would be for the least toxic of those available. An IPM program would maintain pest control, but reduce pollution of other agricultural re-sources and minimize detrimental effects on environmental quality. Bene-ficial organisms would be less likely to be destroyed and the induction of resistance in pests would be minimal.

Growth regulators and mechanization would be used in sustainable agri-culture to reduce labor needs. However, in comparison with traditional prac-tices, the tendency would be toward a conservative application of mechaniza-tion and search for more efficient machinery and practices to reduce energy consumption. The thrust would be to maintain or improve agricultural pro-ductivity while ever mindful of labor input and energy consumption.

### Organic agriculture has much in common with sustainable agriculture.

The same stress is placed upon the use of renewable resources, the need for conservation of resources (energy, soil, and water), and the maintenance of environmental quality. Organic farmers have no desire to eliminate mecha-nization. They are every bit as aware of productivity and economics as the conventional farmer.

The one major difference is in the use of agricultural chemicals. Sustaina-ble agriculturists think in the sense of supplementation; organic agriculturists think in the sense of replacement. Fertilizers synthesized from chemicals are avoided. Nutrient requirements are met solely with animal and green ma-nures, legumes, organic wastes, and mineral-bearing rock dusts. Pesticides are replaced by mechanical cultivation, mulches, biological controls, and other forms of natural control.

The problem of distinguishing between the two agricultural systems arises with borderline cases. Some organic farmers resort to the use of agricultural chemicals when organic practices fail or give unsatisfactory results, although their use is very selective and sparing. Studies by a USDA study team on organic farming concluded that organic farmers were spread over a wide spectrum (USDA 1980). The USDA (as well as borderline organic farmers) considered borderline cases to be practitioners of organic farming, even though they departed occasionally from purist practices.

From the preceding discussion it is clear that organic farmers are practi-tioners of sustainable agriculture, but not all sustainable agriculturists are organic farmers. Regardless of their approaches to sustainable agriculture, both are motivated by the following concerns:

- *Energy*-Increased costs, variable availability at critical times, heavy re-liance on, future limitations
- *Agricultural Chemicals*-Increased costs, heavy reliance on, environ-mental damage, health hazards, increased resistance of insects and weeds to
- Soil-Losses of organic matter, decreased productivity and tilth, erosion
- *Water*-Pollution, future limitations
- *Future*-Sustainability of agricultural production and profits

The objectives and practices of sustainable agriculture owe much to organic agriculture. Concerns leading to the concept of sustainability origi-nated with organic farmers and their advocates. In time these "organic" concerns spread into the conventional farming, public, and governmental domains. These concerns became the impetus for a serious reformation of agriculture during the 1970s, a new direction in which much of the organic agriculture philosophy became integrated with the high productivity prac-tices of conventional agriculture. The new merger was deliberately devoid of the excesses of chemical agriculture, essentially a merger of moderation.

Since organic agriculture preceded sustainable agriculture and embodies much of its philosophy, it is productive to analyze the accumulated data of organic agriculture. Many of the practices and conclusions about the value of organic agriculture have been documented over a lengthy time and are applicable to sustainable agriculture. Analyses of organic agriculture include those of Beradi (1978), Eberle and Holland (1978), Lockeretz *et al. (1976,* 1978, 1981), Oelhaf (1978), Parr *et al.* (1983), Power and Bezdicek (1984), Roberts *et al.* (1979), USDA (1978, 1980), and Vail and Rozyne (1978). Emphasis in this chapter is placed on results derived from the combined practices inherent in organic farming. Some of these practices may already be shared by or be applicable to conventional farming. Analyses of individual practices, including actual and/or projected results in the context of the transition of traditional to sustainable agriculture, are presented in the follow-ing chapters.

## Prevalence, Size, Common Practices

No reliable data exist for the actual number of U.S. farmers who practice some or all organic farming. In 1980, the USDA estimated there were at least 11,200 organic farmers if a strict definition was used, and at least 24,000 if farmers who practice organic techniques on part of their cropland are in-cluded. The USDA belief is that the actual number will be much larger when complete documentation becomes available (USDA 1980).

Contrary to popular belief, not all organic farms are small. They range in size up to 567 ha (1400 acres); combined conventional/organic farms of just over 2429 ha (6000 acres) exist. The success of these farmers indicates that size is not a stumbling block (this will be covered later under economics). The majority of organic farms are owner operated; they are located in all regions except the South, where certain factors (climate, soil, insects, market conditions) limit organic agriculture. Organic farmers represent all ages and levels of education and experience (USDA 1980).

Certain cropping and cultural practices are an integral part of manage-ment on an organic farm. Common elements include legume-based rota-tions, green manures, crop residues, cover crops, animal manures, rock dusts, and small amounts of organic fertilizers. Common crops are forages, vegetables, small fruits, and small grains. Tillage is primarily with chisel plow or disk and often is shallow (3-4 in.). Insects are controlled through selective rotations, biological methods, and organic insecticides. Weed control is usu-ally through proper timing of tillage and planting, crop rotations, and mow-ing. Mechanization is not shunned and efforts toward soil and water conser-vation are excellent (USDA 1980).

## Energy Usage

An examination of the energy usage by organic farmers indicates that their operations are more energy efficient than those of conventional farmers. As expected considerable savings are possible by the elimination of agricultural chemicals, especially fertilizers, which are highly energy intensive. This saving of energy is partially offset by increased energy consumption for additional mechanical cultivation and somewhat reduced yields (see next section, Crop Productivity). Documentation, though not extensive, does exist in the liter-ature.

Lockeretz *et al*. (1976, 1978, 1981) conducted intensive research on corn production in the Midwest. Their study spanned 5 years and compared con-ventional vs. organic farming on several parameters. While the farms com-pared were matched as closely as possible in terms of soil, size, and types of crops, they did differ somewhat in the crop mix. The conventional mixed grain-livestock farms had the following crops in descending order of acreage: corn, soybeans, hay, oats, and wheat. The hay and soybeans were reversed for the mixed grain-livestock organic farms because of their rota-tional requirements. About 10-15% more corn or soybeans was grown on the conventional farms, but about 10% less oats and wheat, than on the organic farms. Because of this relative difference in crop mix, the results were aggregated for all the cropland on each farm and the energy value was expressed per unit value of production. The results indicated that the conven-tional farms required 2½ times as much energy per unit value of production than did the organic farms. Most of the difference was attributed to corn.

Berardi (1978) looked at the energy profile for the production of wheat on organic and conventional farms in New York and Pennsylvania. The conven-tional growers required about 43% more energy; however, once an adjust-ment was made for yield differences, the difference in energy use was only 18%. Further studies with wheat indicated net energy savings of 15-47% were achieved by organic farmers, and a 25% energy savings was noted for barley (USDA 1980).

## Crop Productivity

Several problems exist in assessing yield comparisons between organic and conventional farms. First, there is a paucity of yield data from long-term, well-designed experiments. Replicated, paired experimental plots (organic vs. conventional farming) must be examined over long periods in various geo-graphical locations and under many different conditions and practices before valid comparisons can be made. In fact, such experiments are extremely difficult for several reasons:

- Large numbers of variables-many different soils; variations in climate; variations in crops; sources of organic matter/fertilizer; variations with insects and diseases
- Large range of organic and conventional farming practices-tillage varia-tions; plant protection variations; timing differences for planting, fertilizing, and control of pests; rotation practices
- Prolonged length of time and large number of experiments needed to assess system stability and statistical significance of data

The data that do exist are subject to criticism. Some experiments use replicated, small-plot comparisons of individual treatments. These studies can be challenged fairly as an inadequate representation of organic farming as actually practiced, in that they do not reflect the reality of integrated practices with their additive and synergistic effects.

Other experiments are short-term comparisons with large plots, actual fields, or entire farms, but with small numbers of comparisons. Possible criticisms are that substantial errors of undeterminable nature are quite pos-sible due to the difficulty of controlling or matching variables such as soil or climate, and the difficulty of accurate measurement of yields in large areas. Certainly they do not reflect yields indicative of the many possible variations on the national level.

With these limitations in mind, data on yield comparisons can provide some insight. However, these yield data are not necessarily a reliable indica-tor of the long-term performance of organic agriculture, nor are they applica-ble to any organic farm, regardless of variations in crop, soil, and climate conditions.

The most complete comparative study to date of crop yields on organic and conventional farms was by Lockeretz *et al.* (1981), which was discussed in the section on energy conservation. The yield data was based both upon farmers' reports and actual field measurements. These data were derived from neighboring organic and conventional farms of the same soil type, the same crop variety, and the same or close planting time. Yields on organic farms were 8% lower for corn, 5% lower for soybeans, and 43% lower for wheat than on conventional farms. Berardi (1978) also noted a 22% reduc-tion in wheat yields on organic farms.

Sufficient data on corn were available for Lockeretz *et al.* (1978) to con-duct a more detailed examination of yield. When growing c()nditions were better than average, corn yield differences were generally higher than the average 8%. When growing conditions were poorer (e.g., during drought), corn yields on organic farms approached and even exceeded those on con-ventional farms. Yields of corn on three of the 26 organic farms were higher than 10 MT/ha (159 bu/acre). These are good yields for even a conventional farm.

Data on average yields of important crops on organic and conventional farms in the Midwest are shown in Table 1. The years represented were during a period of serious drought in the Midwest which affected yields in the Corn Belt. As observed by Lockeretz *et al.* (1981), the combined data appear to suggest that during periods of drought corn yields on organic farms tend to approach those on conventional farms and that yields of soybeans, wheat, and oats on organic farms can be reasonably competitive with those on conventional farms. The advantage of organic farms during drought un-doubtedly derives from the emphasis placed upon replacing or increasing soil organic matter. Organic matter contains micropores and is noted for increasing the water retention properties of soils.

**Table 8,1. Average Crop Yields on Midwestern Organic and Conventional Farms, 1973-1976**

| | *Bu/acre* | |
|---|---|---|
| *Crop* | *Organic* | *Conventional* |
| Corn | 77.9 ± 5.4 | 80.6 ± 7.6 |
| Hay | 4.8 | 3.7 |
| Oats | 58.3 ± 3.3 | 57.0 ± 4.7 |
| Soybeans | 30.0 ± 2.9 | 29.9 ±4.0 |
| Wheat | 31.4 ± 3.8 | 34.4 ± 4.1 |

*Source*: Klepper *it al.* (1977), Lockeretz *et al.* (1978), and roberts *et al.* (1979)

*a* Each average is based on 6 or 7 yield reports derived from both annual averages of actural comparative experimental farms and couty data. Hay based upon limited data.

*b* Metric equivalents: 16 bu/acre=14 hectoliters/ha, or about 1 MT/ha.

Better research studies are needed, as stated previously, to assess the long-range productivity of organic farms under the variable conditions en-countered throughout the United States. Still it is reasonable to conclude from the available data that yields on organic farms during drought are comparable to or slightly better than those on conventional farms. As such, organic farming practices have potential value in areas of existing dryland farming, or in areas where irrigation is becoming limited by cost or water problems.

Under ideal conditions organic farming yields would vary from moderately less than to nearly comparable with conventional farming yields. Crops heavily dependent upon fertilizers and pesticides-in particular, corn and wheat—would yield somewhat lower on organic farms. The difference with less demanding crops, such as soybeans, would be smaller, and essentially un-noticeable with crops of minimal demand, such as oats and hay. Conventional farmers switching to organic practices would experience larger yield dif-ferences than those reported until soil organic matter and nutrient levels became well established through cumulative organic techniques. Recently established organic farmers confirm this conclusion (USDA 1980).

One caution is needed here. Conclusions about the desirability of organic farming should not be based strictly upon yield. The energy savings dis-cussed previously would more than compensate monetarily for the observed yield decreases. Other factors must be considered, such as current social or political views about organic agriculture in the context of resource sus-tainability and pollution. The effect of these views is difficult to quantitate, but they are likely to become a stronger incentive in the future. Such a force could lead to more organic farms and/or an increased acceptance of some organic methods on conventional farms. An interesting possibility for con-ventional farmers might be the adoption of organic practices emphasizing the maintenance of soil organic matter coupled to decreased use of agri-cultural chemicals. Such a modified farm should experience no reductions in yield, have minimal problems in terms of pollution, and be viewed favorably in the social and political sense.

## Economics

Limited data is available on the economic performance of organic farms. The present consensus is that net returns from crop production can be similar for some organic and conventional crop-livestock farms (USDA 1980). More data will be required before this conclusion can be accepted as generally valid regardless of farm size, type, and all crops. Several individual studies are discussed in this section.

Eberle and Holland (1978) compared the crop returns on three pairs of conventional and organic farms in Washington. The net return per acre was 33% higher on the conventional farms. The data have been questioned because of the small sample size and the difficulty of pairing factors. In another study in the Northwest area (Kraten 1979), six organic farms, some of which used small amounts of chemical fertilizers, showed a 22% higher net return than conventional farms with similar crops. In the western Corn Belt, Roberts *et al.* (1979) compared 15 organic crop-livestock farms against conventional farms. The majority of the organic farms had net re-turns that surpassed those of the conventional farms.

Lockeretz *et at.* (1981) conducted two studies on the economic perfor-mance of organic farms. The first study, from 1974 to 1976, covered 14 organic crop-livestock farms with an average size of 172 ha (425 acres) in five states: Illinois, Iowa, Minnesota, Missouri, and Nebraska. The organic farms were compared with nearby conventional crop-livestock farms of similar size and soil type. The second study included 23 organic farms in 1977; 19 of these were repeated in 1978. In this study the organic farms were compared with a conventional farm profile determined from data derived from county and federal sources. The organic farms in the second study were located in Illinois, Iowa, and Minnesota and averaged 95 ha (235 acres); their livestock was restricted to beef and/or hogs and their soil had been mapped. Conventional farms of a similar nature in the same localities had a similar average size (96 ha).

The data collected included rates of seeding, choice or cultivars, fertilizer applications, manure and other applications, field operations (tillage, cultiva-tion, harvesting), and yields. The results were expressed as cropland aver-ages and covered the following: cultivated crops, crops used in meadow rotations and for soil improvement, and crops that failed. Market values of the crops were determined based upon reported yields and the state market price, regardless of the ultimate use of the crop. Some crops were sold and others used for livestock. The premium prices offered for some organic crops were not reflected in the calculated values. Fixed or overhead operat-ing expenses were assumed to be the same for the organic and conventional farms, based upon their similarity in size and machinery. Therefore, only direct or variable operating expenses were included: seeds, materials, fuel, labor, repair of equipment, and drying of crops. Labor charges were calcu-lated at prevailing rates, even when work was performed by the farm family.

From 1974 to 1977 the average net returns of organic and conventional farms were within 4% of each other. Organic farms had a lower market value by 6-13%, but their operating costs were less by a similar amount. However, in 1978 a change occurred. Although the operating costs for the organic farms continued to be lower and the net return per hectare was comparable with that in 1974 to 1977, their income per hectare was 13% below that of the conventional farms. The reason for this was that the average net return per hectare for the conventional farms was 21% higher in 1978 than in 1974-1977. Thus, the organic farms did not perform any worse in 1978 than in the previous study years; rather, the conventional farms performed significantly better.

The only discernible difference in these study years was that the severe drought conditions, which existed in 1974-1977, were ameliorated consid-erably in 1978. During the drought years (1974-1977) the yield benefits of the agricultural chemicals used on the conventional farms were minimal because other factors (primarily soil moisture) limited yields. However, in 1978 when growing conditions were more favorable, conventional farms appeared to have a competitive advantage over organic farms, as noted previously. Additional studies are needed to resolve this point. Nonetheless, sustainable agriculture, which employs both organic and conservative con-ventional practices, should not be at a disadvantage during periods of favor-able growing conditions.

The USDA (1980) formulated Midwest farm budgets based upon available data in order to compare economically crop rotations on organic farms and continuous conventional crop practices. The analysis assumed that yields on organic farms were 10% lower. Rotations tie up part of the cropland with forage legumes, such as alfalfa; on conventional farms this land would be producing either corn or soybeans. Since corn and soybeans command a higher price, potential income is reduced in proportion to the amount of land tied up in forage legumes.

Three rotations were compared with conventional continuous corn and soybeans in the USDA study. The conventional system produced 21.6% more income than the 7-year rotation (alfalfa, alfalfa, corn, soybeans, corn, soybeans, and oats), 24.6% more income that the 4-year rotation (alfalfa, corn, soybeans, oats), and 30.4% more income than the 5-year rotation (alfalfa, alfalfa, corn, soybeans, oats). In assessing these results, some factors must be considered. For example, in the 7-year rotation corn and soybeans are produced on only 57% of the total acreage annually. Yet the organic farm yielded an income not 57% but 82% that of the conventional farm with continuous corn and soybeans. Energy savings on the organic farm helped reduce the income difference between the two systems. In the long run the profit difference could decrease and even reverse if the price of energy increases substantially and the yields on conventional farms decrease as soil resources dwindle. The organic farmer is essentially turning part of his poten-tial income into renewal of the soil (by adding organic matter) in order to assure sustainability of future crop production. The conventional system maximizes present income and is not as concerned about viewing soil as a long-term investment. In addition, rotations could become increasingly cost effective if demand for forage crops becomes stronger.

Projections about the economic impact of shifting from conventional to organic farming have been made by the USDA (1980) and Olson and Hardy (1979). Some interesting conclusions were reached. Small farms at the time of the studies accounted for 69% of the total farms and had annual sales of less than $20,000. These farms brought in 11 % of total farm cash receipts. All these small farms could shift to organic farming with little economic impact upon the U.S. economy; however, a total shift to organic farming would have a major impact on the U.S. economy.

If agriculture shifted entirely to organic practices, crop production would still meet U.S. needs, but surplus for exports would be essentially eliminated. Food prices would rise significantly. The cost of the shift would be spread over 3-5 years, the time required for a shift from organic to conventional farming. The effect of a more gradual increase of organic farming in the future is difficult to assess based upon the present inadequate data base. Major savings in energy through decreased use of agricultural chemicals would only be possible if the large conventional mixed crop-livestock farms and the large farms producing the major crops switched to organic practices.

## Labour

A number of variables determine the intensity of labor required in a farm operation. These include the types of crops and livestock, soil topography and type, the size and type of equipment and machinery, type of agricultural practice, and the efficiency of the management and labor. As might be expected, labor intensity varies from organic farm to organic farm and from conventional farm to conventional farm. This variation makes it somewhat difficult to make a general assessment of the comparative labor differences between organic and conventional farms.

Labor usage on organic and conventional farms is shown in Table 2.2. Although information is limited, it appears that organic farms require more labor for management than conventional farms. The increased labor can be attributed to the control of insects, diseases and weeds through mechanical and natural methods as opposed to chemical pesticides. The difference can be kept to a minimum, if hand weeding or hand picking of insects is not used. Such practices were responsible for a 300% difference in labor require-ments of two organic grain farms in Washington (Table 8.2). The use of horses instead of powered vehicles also increases labor requirements (Table 8.2). On comparable organic and conventional farms with similar mechanization and production, the labor input on organic farms, according to Lockeretz *et al.* (1981), was 12% more per unit value of crop production, or 3% more per unit value of land. They concluded that differences in labor reflected the basic difference in cultivation and crop mixes (less corn and soybeans, more forage legumes) practiced by organic farmers.

**Table 8.2. Labor Requirements on Organic and Conventional Farms**

| *Crop* | *Location* | *Organic* | *Conventional* |
|---|---|---|---|
| Corn, soybeans, small grains | Corn Belt | 8.2(3.3) | 7.9(3.2) |
| Corn, soybeans, small grains | Corn Belt | 7.4(3.0) | 6.9(2.8) |
| Grain | Washington | *14.0(5.7)a* | 1.5(0.6) |
| | | 4.7(1.9) | 3.2(1.3) |
| Wheat | New York, | 21.0(8.5) | 8.9(3.6) |
| | Pennsylvania | *13.1(5.3)b* | |

*Source*: Lockeretz *el al.* (1976), Klepper *el al.* (1977), Eberle and Holland (1978), and Berardi (1978).

*a* Insects controlled by hand labor.

*b* Same group but excluding organic farmers using horses in place of mechanization.

Additional labor information in different forms is also available. Klepper *et al.* (1977) showed that Corn Belt organic farms required 19.8 *hrl* $1000 of corn, soybeans and small grains produced as opposed to 17.8 hr for conven-tional farms. Lockeretz *et al.* (1976) in a similar study for the same crops gave a figure of 0.06 *hr/bu* and 0.05 *hr/bu* for organic and conventional farms, respectively. Oelhaf (1978) showed that most vegetables and fruits required more production labor when produced organically than when pro-duced conventionally. The only exception observed by Oelhaf (1978) was intensive tomato production for the fresh market, which required similar labor inputs for organic and conventional systems. Another exception was observed by Roberts *et al.* (1979), who examined labor costs of organic and conventional western Corn Belt farms. They concluded that labor costs for corn, oats, and wheat were less in most cases on organic farms, but more for soybeans.

The need for additional labor on most organic farms does pose some problems. Good farm labor cannot always be found readily, as indicated by a study of organic farms in Maine by Vail and Rozyne (1978). The USDA (1980) feels that labor is a major limitation to increasing the size of some organic farms, especially for vegetable crops requiring hand weeding, and for increasing the number of farms shifting from conventional to organic farming.

## Conservation and Environmental Quality

The idea of conservation and maintenance of environmental quality is inherent in the agricultural practices utilized by organic agriculturists. In prac-tice organic farmers rely heavily upon crop and soil management practices that aid water infiltration, resist soil erosion, improve soil tilth and productivity, recycle organic wastes, and reduce pollution of the soil and water.

## Soil

Conventional agriculture is exacting a heavy price in order to achieve its high rates of crop productivity. The price is "mining of the soil," or soil erosion. The system can only work so long under these conditions. Present estimates are that soil erosion is leading to a drop in crop productivity equiv-alent to an annual loss of 506,072 ha (1.25 million acres).

The organic farmer uses a number of practices that are extremely effective in reducing soil erosion. Continuous cropping is replaced by rotations that include meadow, legumes, and small grains along with row crops. This can result in a reduction of soil erosion. Cover crops, green manure crops, and reduced tillage help to conserve surface crop residues and to increase water infiltration. Cover crops also help prevent soil losses during the unproductive part of the growing season. Maintenance or increases of soil organic matter by application of manures and other organic wastes increases water infiltra-tion and storage. These practices, therefore, decrease surface water runoff and wind blowoff, thus reducing losses of soil through erosion. Increased water storage and less runoff also reduce the contamination of water sup-plies with nutrients and pesticides (if they are used at all).

The effectiveness of these practices can be evaluated mathematically through the use of the C factor (cover and management factor) in the Univer-sal Soil Loss Equation (Wischmeier and Smith 1978). Values for the C factor range from 0.001 for well-managed woodlands to 1.0 for continuous fallow; these values are directly proportional to differences in soil loss caused by differences in the management of soil and crops.

USDA (1980) analyses of organic farming practices have demonstrated their value for erosion control. For example, where rotations utilizing sod crops (grass and legumes) constitute 25-40% of the crops, soil erosion is 33-12.5% of that observed with continuous row crops and conventional tillage. Cover crops, when they follow harvested crops such as silage corn, potatoes, and most vegetables, are important in protecting the soil, since the preceding harvested crops leave little residue. Cover crops in such systems can reduce soil losses through erosion by 50%. The use of the chisel plow and disk by organic farmers, instead of the moldboard plow, can reduce soil erosion by 20-75%. The effectiveness of this practice results from the place-ment of crop residue at or near the surface.

Increased organic matter in the soil can also reduce the loss of soil. On the average, an increase of 1 % in soil organic matter can decrease the potential for soil erosion by 10%. Such increases in organic matter are possible with continuous organic practices of manure applications and grass-legume ro-tations (Cooke 1977). Other soil conservation practices used by organic farmers include contour planting and plowing, terraces, and grassed water-ways. The quantitative value of these practices was not evaluated in the USDA study of 1980.

The collective value of these and other practices has not been researched to any significant degree in reliable, comparative studies of organic and conventional farms. Lockeretz *et al.* (1981) estimated that organic crop-livestock farms in the Corn Belt had a loss of soil by water erosion about one-third less than that of conventional paired farms. This estimate was based on rotations and did not include the reduced tillage practices of the organic farmer, which would have increased the difference even more.

Besides preventing soil erosion, the preceding practices of organic farm-ers are the very management practices that have been advocated by the USDA and Land Grant universities for improving soil tilth and productivity (USDA 1980). Over a long period of time this type of agricultural manage-ment can increase aeration, aggregation, permeability, and water storage capacity; decrease compaction and crusting; and improve levels of beneficial soil organisms (insects, earthworms, and microorganisms).

## Water

A number of organic practices increase water infiltration, thus increasing water availability over the short term (USDA 1980). This improvement arises from increases in surface crop residues provided after plowing out or under cover crops, green manures, and sod crops, and from reduced tillage prac-tices. The maintenance and increase of soil organic matter also aids water infiltration. The effectiveness of these practices was evident in the findings of Lockeretz *et al.* (1981) that the drought in the Corn Belt during the mid-1970s caused yield reductions for conventional farmers but not for organic farmers.

Increased water-holding capacity can result from long-term increases in or maintenance of a high level of soil organic matter. Reliable assessment of the results from short-term changes in organic matter is difficult (USDA 1980). Soils most likely to show increased water retention from the addition of organic matter are sandy soils. Further studies are needed on the long-term advantages of organic farming practices in terms of soil organic matter and its effect on crop productivity.

## Environment

Organic farming practices that reduce soil erosion also reduce surface water runoff and thus are less likely than conventional practices to result in contamination of water supplies with nutrients and pesticides. The organic practice of not using agricultural chemicals also reduces the environmental impact of farm operations.

Excessive use of chemical fertilizers can lead to pollution of groundwater with nitrogen and phosphorus, especially if associated with high rainfall or irrigation, shallow-rooted crops, and sandy soils. Organic farmers use little, if any, chemical fertilizers. Instead they rely upon recycling of nutrients in their management approach; these nutrients generally are not susceptible to leaching, even when applied excessively. Indeed, many organic farmers tend to operate at nutrient levels below the maximum needed for optimal yields (USDA 1980). Less nutrients and less leaching coupled with practices that reduce soil erosion by water runoff minimize nutrient flow from the farm to groundwater.

The effectiveness of organic practices in reducing nutrient flow from farm-land can be attributed to the slow release of nitrogen from organic materials such as manures, compost, sewage sludge, and residues from cover crops, green manures, and rotations. The continual crop cover provided by rota-tions, cover crops, and green manures utilizes the released nitrate, leaving little for leaching away. In addition, the rotation of crops that require high amounts of nitrogen (e.g., corn) with crops that require little nitrogen (e.g., soybeans and alfalfa) helps to lower the average amount of nitrate available for leaching. Alfalfa has deep roots, enabling it to retrieve nitrogen past the root zone of other crops. Winter cover crops extract nitrate and soil water, thus reducing the usual loss of nitrates in winter when the land normally lies fallow.

Several studies have demonstrated that sod and cover crops do, indeed, reduce the chance of pollution from nutrient runoff. For example, losses of nitrates and phosphates were 300-600% less with a corn-wheat-clover rotation than with continuous corn (Stewart *et al.* 1975). The planting of corn in a ryegrass cover also resulted in a 50% decrease in runoff, as well as a 40-fold reduction in the loss of soil (as sediment) and of nitrate and phosphate (Stewart *et al.* 1975).

This discussion should not be construed to mean that no threat of pollu-tion exists from organic techniques. Improper management of manures and sludge's can cause problems. Incorrect storage of manures and sludge can lead to pollution and nutrient waste through leaching action. Improper ap-plication or use of contaminated sludges can lead to pollution of soil and water, and possible release of toxic compounds that can end up in the food chain. Excessive use of organic wastes can contribute to pollution of ground-water both during the growing season and after harvest. The excess nitrate not utilized during growth can be leached, and the slow release of nitrate after harvest continues to cause further problems. Runoff removal of some nu-trients is possible if manures and sludges are applied to frozen or snow-covered fields.

In addition, adequate phosphorus (P) and potassium (K) must be applied to balance the nitrogen (N) added from either organic or chemical sources. In one study, much nitrate remained in the soil after harvest of wheat in a wheat-corn rotation fertilized with 120 kg N/ha (107 Ib N/acre) when no P and K were added. In contrast, little N was found in the soil at depths up to 2 m when P and K were added at rates near 25 kg/ha (22.2 Ib/acre). This problem should concern organic farmers, since the USDA (1980) study found that few of the surveyed Midwest organic farmers attempted to balance the nutrient budget with phosphorus or potassium. Lockeretz *et al. (1976)* also noted an average net deficit of 13.4 and 45.9 kg/ha (12 and 41Ib/acre) of $P_2 0_S$ and $K_2 0$, respectively, on organic farms in the Midwest.

Even larger accumulations of soil nitrate are quite possible with organic farming. For example, if a legume crop followed a wheat harvest, as would be likely in a typical rotation, the concentration of nitrate would rise because of nitrate fixation by the legume. The potential for nitrate contamination of groundwater would exist. Under these conditions a non leguminous crop would be better, since oats, timothy, and rye can reduce leaching losses 40-60% (Stewart *et al. 1976).*

Pesticide pollution by organic farmers is minimal because they use non-chemical methods of pest control most of the time. Even if occasional lapses occur, their effective practices to reduce soil erosion and, hence, runoff would minimize the problem of pesticide contamination of groundwater. Further, their choice of pesticide would likely be the least toxic and most biodegradable.

## Future Prospects

The available research data and the continued existence of organic agriculture in the face of economic reality suggest that at least this particular form of sustainable agriculture is viable. It is unlikely, however, that organic farming in its pure form should be or in the future will be the predominant form of sustainable agriculture in the United States. There are several rea-sons for this viewpoint.

Arguing against the widespread adoption of organic farming are the un-certainity of labor supplies, the probable loss of agricultural exports, and the reluctance of many conventional farmers to give up agricultural chemicals. On the other hand, the organic approach has much to offer the conventional farmer who wishes to practice sustainable agriculture. First, a definite reduc-tion in energy costs accompanies use of organic practices. Conservation of agricultural resources, especially soil, also is assured. Maintenance of en-vironmental quality can be achieved, even with some use of agricultural chemicals. Finally, reasonable productivity can be attained with organic practices.

The ideal solution would be an integration of the best from both organic and conventional farming practices, a new direction designed to sustain agriculture. Essentially, organic practices would be retained, but supple-mented with judicious use of agricultural chemicals. Integrated pest manage-ment approaches would be used to control pests. Soil maintenance would involve blends of organic and conventional techniques designed to maintain or increase soil organic matter and fertility. The thrust would be to minimize energy usage and pollution, but to maximize conservation of resources. The extent and proportions of combined organic and conventional farming prac-tices would be determined by personal belief, the desired levels of economic return, and the desired investment in agriculture's future.

## Research Needs

Although much is known about organic farming, extensive research is needed to assess the full potential of this farming system as a form of or source of practices for sustainable agriculture. Research needs have been thoroughly established by the USDA (1980) and prioritized by Parr *et al.* (1983).

One area of immediate urgency is the investigation of organic farming not as a collection of individual practices, but as a collective practice, that is, a holistic research approach. The practices of organic farmers need to be examined for their interrelationships. For example, what are the interre-lationships between organic waste recycling, energy conservation, nutrient levels and availability, and energy conservation? What are the effects of organic wastes coupled with rotations, cover crops, and the lack of agri-cultural chemicals upon the microbiology and biochemistry of the soil? How does organic agriculture compare with conventional agriculture in terms of economic return and several resource parameters, such as soil and water sustainability, environmental quality, energy conservation, and crop produc-tivity?

Much field research over long periods of time will be required. Research must be done through an interdisciplinary approach, if the holistic examina-tion is to succeed. Various experts will be needed in the areas of agronomy, biochemistry, climatology, ecology, economics, entomology, microbiology, pathology, physiology, and soil science.

In addition to the gathering of data, there will be a need for interpretation of the results and an assessment of what effect changes in practices have upon organic farming as a whole, but especially upon sustainability and environ-mental quality. Here the use of computer-assisted mathematical modeling will be mandated.

This will require the application of systems science to organic agriculture, that is, the resolving of organic farming into various components and sub-systems. Each component will become a mathematical model, able to be described individually and in relation to the other models. The collective interaction of the models becomes the system, or a working simulation of organic farming. With this approach the models can be manipulated to predict the system's behavior when given different inputs. Systems science has been applied successfully in other areas of agriculture, such as inte-grated pest management (see Chapter 9). However, much work remains, if system science is to be applied to organic agriculture.

Many other areas require further study. For example, decreased yields are observed for 3-4 years during the conversion from conventional to organic farming. The causes, beyond weed problems, need to be identified in order to improve yields and economics during the transitional period. This aspect would appear to be especially important if conventional farmers are to be encouraged to become organic farmers, or to adopt specific organic practices.

A number of questions remain unanswered in terms of organic wastes. The inventory of organic wastes is presently incomplete, especially for indus-trial wastes produced by food processors, logging and wood product manu-facturers, pharmaceutical producers, leather processors, and so forth. Data on all wastes are needed in terms of kinds, amounts, availability, and nutrient content.

Improved methods for handling each organic waste need to be developed. Is processing before land application necessary to reduce phytotoxicity or health hazards, or to enhance nutrient availability? Do any environmental or other constraints apply to long-term use of organic wastes? Can agricultural uses be improved in terms of labor, nutrient recovery, and economics? What are the effects of organic wastes upon soil properties, such as maintenance of organic matter, erosion control, tilth, water infiltration and retention, levels and activities of beneficial soil organisms, and fertility? Effects upon fertility are especially important; what is the relationship between application rates and frequency and nutrient availability? Soil tests will need to be improved considerably if these questions are to be answered.

Can the benefits resulting from application of organic wastes to soils be increased further? For example, there appears to be a synergistic effect when organic wastes are combined with chemical fertilizers. What are the com-binations that produce optimal effects upon crop yields?

The relationship between energy conservation and use of organic wastes needs to be analyzed. The application of organic wastes to soils offers some energy savings, since fertilizer needs are reduced, and produces benefits resulting from maintenance of organic matter, such as erosion control. If some of these wastes were diverted to direct energy production, would the gain in energy offset the loss of benefits derived from use of wastes in the soil? What percentage can be diverted without harm to long-range soil pro-ductivity? What must be done to maintain soil productivity on biomass plantations?

Other areas need investigation. Can biological nitrogen fixation by legumes be improved, perhaps through selective breeding? Can it be enhanced by any changes in cultural practices? Can better non chemical ways of controlling insects, diseases, and weeds be developed? Can plant breeders produce crops that are better adapted to organic farming?

Of course, results from all these avenues of research must reach the mainstream of agricultural activity, including education, extension, farmers, and the public. This is especially important for successful conversion or partial adaptation. Dissemination of research results must be improved sub-stantially; organic farming practices have been poorly communicated and understood in the past.

One final thought. The amount of research suggested here is extensive and perhaps easily dismissed because it seems to only apply to organic farming. Although such research would certainly improve organic farming practices and perhaps increase the number of successful conversions, its most important practical outcome would be development of practices tai-lored specifically to improve the sustainability of conventional agriculture.

## REFERENCES

Berardi,, G. M. 1978. Organic and conventional wheat production: Examination of energy and economics. Agro-Ecosystems 4, 367-376.

Cooke, G. W. 1977. The roles of organic manures and organic matter in managing soils for higher crop yields: A review of the experimental evidence. *In* Proc. Intern. Seminar on Soil Environment and Fertility Management in Intensive Agriculture, Tokyo, Japan. Ministry of Agriculture, London.

Eberle,, P. and Holland, D. 1978. Comparing organic and conventional grain farms in Washington. Tilth (Spring), 34-35.

Klepper, R., Lockertz, W., Commoner, B., Gertler, M., Fast, S., O'Leary, D. and Blobaum, R. 1977.

Economic performance and energy intensiveness on organic and conventional farms in the Corn Belt: A preliminary comparison. Amer. Jr. Agric. Econ. 59, 1-12.

Kraten,S. L. 1979. A preliminary examination of the economic performance and energy intensiveness of organic and conventional small grain farms in the Northwest. M.S. Thesis. Washington State University, Pullman.

Lockeretz, W., Klepper R., Commoner B., Gertler, M., FAST, S. and O'Leart, D. 1976. Organic and Conventional Crop Production in the Corn Belt: a comparison of Economic Performance and Energy Use for Selected Farms. Rept. CBNS-AE-7. Center for the Biology of Natural Systems, Washington Univ., St. Louis, MO.

Lockeretz, W., Shearer, G., Klepper, R. and Sweeney, S. 1978. Field crop produc-tion on organic farms in the Midwest. J. Soil and Water Cons. 33, 130-134.

Lockertz, W., Shearer, G. and Kohl, D. H. 1981. Organic farming in the Corn Belt. Science 211,540-547.

Oelhaf, R. C. 1978. Organic Agriculture: Economic and Ecological Comparisons with Conventional Methods. Halsted Press, New York.

Olson, K. D. and HARDY, E. O. 1979. A National Model of Agricultural Production, Land Use, Export Potential, and Income Under Conventional and Organic Farming Nternatives. Center for Agricultural and Rural Development, Iowa State Univ., Ames. (Unpublished.)

Parr, J. F., Papendick, R. I. and Youngberg, I. G. 1983, Organic farming in the United States: Principles and perspectives. Agro-Ecosystems 8, 183-201.

Power, J. and Bezdicek, D. (Editors). 1984 Organic Farming: Current Technology and Its Role in a Sustainable Agriculture. Amer. Soc. Agron., Madison, WI.

Robertz, K. J., Warnken P. F. and Schneeberger, K. C. 1979. The economics of organic crop production in the western Corn Belt. Agric. Econ. Paper 1979-6. Dept. Agric. Econ., Univ. of Missouri, Columbia.

Stewart, B. A., Woolhiser, D. A, Wischmeir, W. H., Caro, J. H. and Free, M. H. 1975. Control of Water Pollution from Cropland. Vols. I and II. U.S. Dept. Agric. and Environ. Protection Agency, Washington, DC.

USDA 1978. A Bibliography for Small and Organic Farmers, 1920 to 1978. U.S. Dept. Agric., Washington, DC.

USDA 1980. Report and Recommendations on Organic Farming. U.S. Dept. Agric., Wash-ington, DC.

VNL, D. and Rpzume,, M. 1978. Appropriate technology for small farms: Lessons from 31 organic farms in Maine. Idea Paper 11, Main Small Farm Management and Technology Project. Dept. Econ., Bowdoin College, Brunswick, ME.

Wischmeier, W. H. and Smith, D. D. 1978. Predicting Rainfall Erosion Losses: A Guide to Conservation Planning. USDA Agric. Handbook 537.

# 9

# Conservation of Crop Energy

Today the food-producing capacity of American agriculture is unsur-passed by that of any foreign power. The efforts of farmers and agricultural scientists over the last 40 years are responsible for this great achievement. Since 1950 these efforts have resulted in a doubling of productivity and a halving of agricultural labor (Price 1981). Much of the increase can be at-tributed to increased use of fertilizers and advances in pest control, plant and animal breeding, and mechanization. With the exception of results based upon plant and animal genetics, most of the efforts that increased productivi-ty were energy dependent. The price of doubled agricultural productivity is high: a four-fold increase of energy input (Price 1981).

Estimates place the energy consumption of the food and fiber system at just over 20% of the total 1981 energy needs in the United States (Price 1981; Vilstrup 1981). The agricultural segment of the food and fiber system requires at most one-sixth of this system's energy: The remainder is needed for food processing and distribution. The latter includes energy to transport food from the farm to the processor and eventually to the super-market, and energy to process, package, market, store, and prepare the food.

In comparison with 1978 data, energy consumption for field machinery, crop drying, and miscellaneous activities remained about the same in 1980. However, a decrease of 34 and 8.8% occurred for irrigation and transportation, respectively, and an increase of 14% for fertilizers and pesticides, due mainly to increased fertilizer use. The decrease in energy usage attributed to transportation probably resulted from conservation measures and improved fuel efficiency of newer vehicles. The decrease in energy consumption for irrigation probably resulted largely from improvements in irrigation efficiency and some loss of irrigated acreage to dryland farming, a result of energy costs becoming too high for profitable operation of irrigated land in some areas.

At first glance appears to that crops consume most of the agricultural energy input. However, a significant part of the corn, alfalfa, and soybeans is fed to livestock. Animal husbandry, including operations and feed requirements, is thus the main energy consumer in American agriculture. This has led some to question whether livestock production should have such a substantial role in American agriculture.

The energy value of crops before they are processed or fed to livestock ex-ceeds the energy input required for their production. This favorable energy balance changes considerably once the energy input for food processing and distribution is added, or the energy value of livestock products is weighed against the direct consumption of feeds used to raise livestock. This negative energy deficit has been accepted in the past, since it rested on inexpensive and readily available energy and produced high-quality food products that were easily available and convenient.

Recently, some analysts have begun to examine agriculture strictly on an energy basis. All food and fiber products are thought of in terms of energy output and input, as was done in the preceding paragraph. Others look at agricultural data strictly in the economic sense. The cost of energy to pro-duce American agricultural products on the farm is about 10% of their value (Lockeretz 1982). In this view, the food needs of America are satisfied at a reasonable cost to the consumer, and excess food remains for export. In-deed, the value of agricultural exports amount to somewhat over half the cost of imported oil, yet their production on the farm only consumes approx-imately 1 % of our national energy budget. This group feels less alarm about energy input to agriculture and gives it only that attention that economic trends foretell.

Still others think that agricultural products may someday reduce our need to import energy. They point to ethanol from corn, diesel-like fuel from sunflower oil, and petroleum feedstocks from certain oil-rich succulents cultivated on arid lands. Here two factors must be watched: energy input vs. energy output and competition between crops for food or energy.

Regardless of which viewpoint one takes, energy use is a common consid-eration in all cases. As such, energy conservation is an integral part of sus-tainable agricul-ture. However, the holistic view of sustainable agriculture should be kept in mind: Energy conservation must be accorded equal status with conservation of soil and water resources. Resource conservation must also be balanced against maintenance of maximal agricultural productivity with minimal labor input, while avoiding environmental pollution.

**Energetics of Crop Production**

An examination of the literature shows there is much variation in the energy needed to produce various crops. Factors involved in the wide range of energy inputs include the type of crop, the region, and the manner in which it is produced. Price (1981), for example, reported the following energy requirements, expressed in Btu per pound: tobacco, 30,000; feed grains, 2000; cotton, 24,000; and corn, 1200-8400. The energy input for corn depends on the intensity of the production system and the region. (For conversion purposes, 1000 Btu/lb equals 2.32 MJ/kg.)

These numbers must be put into proper perspective for purposes of eval-uating the need for energy conservation in agriculture. Most studies estimate that energy used in food production on the farm accounts for about 3% of total national energy consumption; the highest cited value in the literature is near 5% . Energy consumption for crop production is rather low compared to that for other major activities, such as food processing and distribution at about 17% (Price 1981; Vilstrup 1981).

However, the dependence of crop production upon energy is much more critical than numbers would appear to indicate. A shortfall of energy at a critical junction, such as during periods of pesticide application or during crop harvest, could spell disaster on a national scale. Timeliness of the fuel supply is extremely important. Unfortunately, the timing of fuel requirements is itself a variable that depends upon factors such as crop development, weather, and appearance of pests.

One obvious way to reduce the potential danger of temporary energy shortfalls and to cut production costs is to decrease the total energy input. However, before this can be done, several questions require answers. What crops are energy intensive? Are the energy-demanding crops a significant part of agricultural production, such that energy conservation would produce a tangible result? How is energy partitioned during crop production; that is, could one activity in crop production be made more energy efficient? Are some cropping systems more energy efficient than others? These and other questions can only be answered after an energy audit is conducted for the various crops.

Agricultural scientists have conducted energy audits for the major U.S. crops (Pimentel *et al.* 1973; Heichel 1978). These audits are essentially an accounting of the partitioning of energy resources in crop production. Some of the components of these audits reflect direct energy inputs; these include the fuel needed to run farm structures and various machinery. Other compo-nents reflect indirect energy inputs, for example, the energy needed to make fertilizers, various agricultural chemicals, farm machinery, and buildings.

An excellent source of indirect and direct energy values for crop produc-tion is available (Pimentel 1980). Some examples of interest follow. The total embodied, fabrication and repair parts energy during the reliable life spans of a two-wheel-drive 130-hp tractor (6078 kg or 13,400 lb) and a 4.27-m (14-ft) chisel plow (2213 kg or 5100 lb) is 99,655,341 and 42,696,767 kcal, respec-tively. If the life spans of the tractor and plow are 12 and 15 years, respec-tively, and they are used on a 200-ha (494-acre) farm, the annual combined energy cost for both is 55,755 kcal/ha (22,573 kcal/acre). Energy inputs for farm-delivered fertilizers range from 12 to 15 Mcal/kg N, 0.6 to 1.5 Mcal/kg$P_2O_5$, and 0.5 Mcal/kg$K_2O$. The energy input for crushed and ground limestone is 300 kcal/kg. The average energy input for pesticides is 61,470-86,910 kcal/kg active ingredient for insecticides; 62,770-99,910 kcal/kg for herbicides; and 27,770-64,910 kcal/kg for fungicides.

**Table 9.1. Acreage and Fossil Energy Flux of Major Crops in the United States**

| Crop | *Acreage* [a] *(1000 acres)* | *Fossil energy flux*[b] *(Meal/acre-day)* | *Relative energy rank* |
|---|---|---|---|
| Corn | | | |
| silage | 9,261 | 17 | 1[d] |
| grain | 70,061 | 17 | |
| Soybeans | 67,856 | 5 | 2 |
| Alfalfa | 26,269 | 6 | 6 |
| Wheat (spring) | 19,479 | 5 | 7 |
| Cotton (lint) | 13,214 | 14 | 5 |
| Sorghum (grain) | 12,722 | 16 | 4 |
| Oats | 8,640 | 4 | 10 |
| Barley | 7,233 | 6 | 9 |
| Vegetables and fruit | 6,044.8 | 56[c] | 3 |
| Rice | 3,295 | 34 | 6 |
| Peanuts (for nuts) | 1,398.8 | 18 | 11 |
| Sugar beets | 1,187.2 | 11 | 12 |
| Sugar cane | 735.6 | 9 | 13 |
| Tobacco (flue-cured) | 511.8 | 125 | 8 |
| Pasture | — | | — |
| fertilized | — | 2 | — |
| fertilized, irrigated | — | 23 | — |
| Range | — | 0.03 | — |

[a] USDA (1982).
[b] Heichel (1977, 1978); Klopatek (1978); USDA/FEA (1976). [c]Average value.
[d] Combined silage and grain.

These and other energy inputs for crop production have been summarized for each crop, and then the values proportionately adjusted to compensate for differences in days to maturity from seed or transplants. The final value gives an indication of the average daily rate of energy use per acre in produc-ing a specific crop. The units of this fossil energy flux (FEF) are Mcal/acre-day. Values of FEF for the major crops and their relative importance in the consumption of agricultural energy are shown in Table 1.

Some interesting conclusions can be drawn from Table 9.1. Certain crops require much less energy for production than others. The least energy intensive crop is native range, used by grazing animals as forage. The only management is occasional applications of herbicide to control brush. Small amounts of energy are used to manufacture and apply the herbicide. At the other extreme tobacco requires nearly 4200 times as much energy as native range. Much of this energy consumption occurs in the curing and drying required shortly after harvest. Heavy use of liquified petroleum gas is needed to produce flue-cured tobacco. Fortunately, the acreage devoted to tobacco is small compared to most crops; little potential exists for energy conserva-tion during the flue-curing process, except by eliminating tobacco pro-duction.

Vegetables and fruits have intensive crop systems involving heavy inputs of fertilizer, pest control, and irrigation. As expected, their FEF value is quite high (56), nearly 1900 times greater than that of native range. The input of fertilizer to convert native range to fertilized pasture having managed forage species raises energy consumption 66-fold; the addition of fertilizer and irrigation causes a 767 -fold increase! The energy price for fertilizer, irrigation, and pest control is quite high.

An insight into one way of altering a cropping system to reduce energy consumption can be gained from looking at the values for corn and alfalfa. Silage corn is an annual forage requiring heavy nitrogen fertilization. Alfalfa is also an annual forage, but being a legume, it fixes about 80% of its nitrogen needs from the atmosphere. As shown in Table 1, the FEF of silage corn is 183% greater than that of alfalfa. Heichel (1978) has shown that the different needs for nitrogen fertilizer accounts for much of the energy differential between corn silage and alfalfa. Since leguminous forages after harvest leave nitrogen in the soil for succeeding crops, they can reduce nitrogen fertiliza-tion for nonleguminous crops. Before the ready availability of chemical fertil-izers, farmers used legumes and nonlegumes in rotational crop systems to reduce the need for nitrogen. An old idea has found merit again. The use of this type of rotation is examined later in this chapter in the section on ferti-lizers.

Another approach to energy conservation in crop production is suggested by a comparative study of organic and conventional farms located in the Corn Belt (Lockeretz 1978). Each organic-conventional pair was matched as closely as possible in terms of soil and size. The organic farmers used only organic fertilizers (livestock manure, leguminous green manures, or com-mercial organic fertilizers) and natural forms of pest control; the conventional farmers used conventional techniques, chemical fertilizers, and pesticides. The organic operations produced 11% less market value per hectare, but needed only 40% as much fossil energy per hectare as conventional opera-tions. Much of the energy consumption of conventional farming was at-tributed to the use of nitrogen fertilizers. Efforts to conserve energy by use of organic fertilizers are examined later in this chapter.

The studies of Heichel (1978) and Lockeretz (1978, 1982) clearly show that reducing the use of chemical forms of nitrogen fertilizer can conserve energy. However, their research also pointed out other aspects of agricultural production that consumed considerable energy: farm machinery, tillage, fuel, and pesticides. In decreasing order the three most energy-intensive and clearly defined crop production activities are fertilization, field operations (farm machinery and tillage), and irrigation (Heichel 1978; Waldrop 1981; Lockeretz 1982). Energy considerations associated with these and other crop production activities are examined in detail in the rest of this chapter.

**Fertilization**

A third of the productive capacity of American crops results from the use of nitrogen fertilizers. Estimates are that the United States would require 18 million acres of additional cropland to maintain current production levels if nitrogen fertilizers were restricted to 50 Ib/acre (56.8 kg/ha) (Price 1981).

The production and application of fertilizers accounts for an estimated 30-38% of the total energy input for crop production (Lockerelz 1982; Price 1981). Fuel energy is needed for mining, refining, transporting, and applying fertilizers. Most of the inherent energy of fertilizers is attributed to the natural gas needed to produce nitrogen fertilizers. Average values of inherent energy in fertilizers (not including energy of field application) are 25,000 Btu/lb of nitrogen; 3000 Btu/lb of $P_20_S$ equivalent; and 2000 Btu/lb of $K_20$ equivalent (Frye and Phillips 1981). In terms of gallons of diesel fuel equivalent per pound, the respective values are 0.170, 0.020, and 0.014; in terms of mega--joules per kilogram (MJ/kg), the values are 58, 7, and 4.6, respectively.

Natural gas supplies the hydrogen needed for ammonia compounds, of which 94% are manufactured from gas. About 93% of the nitrogen in fertil-izers is derived from ammonia compounds (anhydrous ammonia, urea, and ammonium nitrate). Anhydrous ammonia is slightly more energy efficient to manufacture than urea and ammonia nitrate. The respective average energy requirements are 46, 69, and 65 MJ/kg. Nitrogen fertilizers consume about 2% of the total natural gas used in the United States (Fluck and Baird 1980).

Fertilization application rates vary considerably depending on soil condi-tions, climate, and crop. Rates on corn, for example, range from 75 to 175 Ib/acre (85 to 199 kg/ha) of nitrogen and from 60 to 150 Ib/acre (68 to 170 kg/ha) each of potassium and phosphorus as $P_2O_5$ and $K_2O$. Nitrogen is the most important nutrient for production of non leguminous crops because it leaches the easiest and tends to be deficient, thus limiting yields more than potassium or phosphorus. Nitrogen fertilizer, being used the most and also the most energy expensive, would seem amenable to more efficient utiliza-tion to save energy.

**Reduction of Fertilizer Usage**

Whatever energy conservation measures are taken to reduce usage of nitrogen fertilizer must be carefully considered against yield. The general relationship of yield to applied nitrogen fertilizer. The rate of increase in the yield of corn (or of any other crop) diminishes per addi-tional increment of fertilizer as the peak yield is approached. However, this still may be economical, if the value of the additional corn is reasonably greater than the cost of the fertilizer. Conversely, any reduction in applied fertilizer, which results in lower levels of utilizable nitrogen, has a price tag: reduced yield. This need not be the case, if the reduction eliminates non-utilizable nitrogen normally lost by leaching or if nitrogen fertilizer is replaced by another less energy-expensive source of nitrogen, such as manure or legumes.

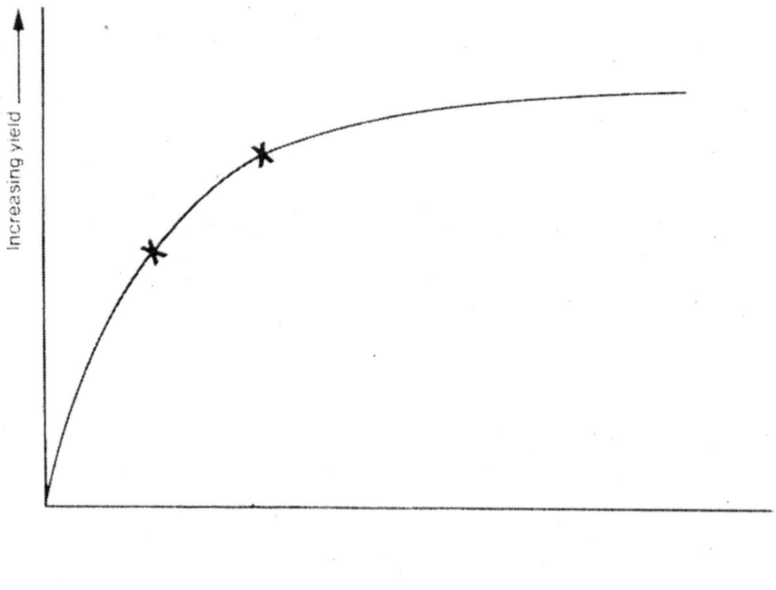


***Fig.* 9.1.** Application rates of nitrogen fertilizer are usually chosen to realize yields in the range indicated by the Xs. Efforts to reduce nitrogen use should not be at the expense of yield, but aimed at eliminating excesses or practices that lead to lost nutrients.

The manufacture of fertilizers could be improved in terms of energy use and ability to use lower-grade feedstocks. Some progress is being made on this front with phosphate nutrients, which are closer than other nutrients to exhausting feedstocks. High-grade Florida phosphate rock is being used up at an alarming rate. Fortunately, research on process improvement and production of phosphate fertilizer from lower-quality ores has been started by the Tennessee Valley Authority's National Fertilizer Development Center (Anon. 1983).

Some approaches to fertilization of cropland can result in decreased usage of chemical fertilizers without causing reductions in crop yield. The amount of fertilizer applied should be based on thorough soil tests prior to initial and supplementary applications of fertilizer. Recommendations of these tests should be closely followed to avoid using too much (or too little) fertilizer.

However, even fertilization rates based on soil test results may be excessive (Liebhardt 1983). Results of soil tests are made on the basis of the best available information, which often is weak in terms of sufficient field calibra-tion data. Also, good tests for available soil nitrogen are lacking, and results appear to be highly variable among laboratories. Many do not take into consideration the amount of nitrogen contributed by a previous legume crop. Often, even when soils test high in fertility, additional fertilizer is recom-mended. Evidence is beginning to suggest that money is now spent on some fertilizer that will not increase crop yields (Olson *et al. 1982).*

Improvement of soil testing, in terms of chemical tests and the data basis upon which results are formulated, appears to be needed. Such improvement, based on present evidence, warrants high research priority. The energy and monetary savings possible through reduced fertilizer usage based on better soil testing could be substantial (Liebhardt 1983).

A good review of the problems with the existing tests for predicting the release of soil nitrogen is available (Barker 1980). Some progress has been made in improving tests for the determination of nitrogen supplied by the soil, for example the method of Fox and Piekielek (1978a). While their test has been shown to be useful under several environmental and soil conditions (Fox and Piekielek 1978b), much research remains. Tests, such as this one, need to be calibrated to crop demands in the field under a wide range of environmental and soil conditions. Although this research will be expensive and time-consuming, the rising costs of nitrogen fertilizer provide consider-ably justification for the effort.

Careful attention should be given to the particular nitrogen fertilizer used and the timing and placement of application (Barker 1980). For example, fall-applied nitrogen is only 37% as effective as side-dressed nitrogen applied when corn is 30-45 cm (12-16 in.) high. Side-dressings are more efficient than spring preplant broadcast treatments. Urea-based fertilizers are more apt to lose nitrogen through volatilization than other types if no rain falls within 3 or more days after application. These same fertilizers also have been shown to produce lower yields than ammonium nitrate with no-till corn. Research is needed to reduce losses associated with urea-based fertilizers.

If fertilizer is applied prior to planting, some leaching or denitrification of nitrogen may occur. Such preplant treatments are therefore somewhat ineffi-cient. For example, corn, which consumes 25% of the total energy con-sumed in U.S. crop production, has only a modest nitrogen requirement during the first 4 weeks after seed is planted (Frye and Phillips 1981). If the application of nitrogen fertilizer were delayed for 4 weeks, it would be reason-able to expect more efficient utilization of nitrogen. This can be the case, if climatic and soil conditions are correct.

Nitrate leaching is the most likely cause of nitrogen loss on well-drained soils. In soils of moderate to poor drainage, greater losses of nitrogen more likely occur through denitrification of nitrate. Increased rainfall accelerates both leaching and denitrification on soils, regardless of drainage conditions. Conventional tillage, as opposed to no-tillage, also increases losses from denitrification and leaching of nitrate.

When nitrogen fertilizers are applied all at planting, the rates for moderately well-drained to poorly drained soil are 50 and 75 Ib/acre (57 and 85 kg/ha), respectively, higher than the rate on well-drained soils. This additional nitro-gen is to correct for greater potential losses on soils with less than well-drained conditions. These rates can be reduced by 35 Ib/acre (40 kg/ha) if up to two-thirds of the nitrogen fertilizer are applied 4 weeks after planting the corn (Frye and Phillips 1981). Such delayed applications appear to be most efficient for no-tillage corn grown on moderately well-drained soil and for conventional tillage corn on moderately to poorly drained soil. Periods of peak nitrogen requirements for various crops are shown in Table 9,2. Depending upon future research, crops other than corn might also be amena-ble to delayed applications of nitrogen fertilizers, with associated reductions in total usage.

**Table 9.2: Time of Peak Nitrogen Requirements of Certain Crops**

| *Crop* | *Peak nitroten demand*<br>*Daysa* | *Stage* |
|---|---|---|
| Field corn | 30-45 | — |
| Sweet corn | 25 | Whorl |
| Cucumber | — | Fruit development |
| Lima beans | — | Flowering and fruiting |
| Radish | — | root formation until 75% complete |
| Southern peas | — | Peak vegetative growth, pod filling |
| Soybeans | — | Peak vegetative growth, pod filling |

*Source:* Based upon data in Barker and Mills (1980). [a]After planting.

### Use of Organic Wastes

Other approaches to reduction of the energy costs involved with use of chemical nitrogen fertilizers include partial to complete substitution by legumes and/or organic wastes. The latter includes manures, crop residues, compost, and sewage sludge. Current USDA estimates are that 90% of the available manures are already applied to land. Even if all the unavailable manures (range droppings) were utilized, manure could supply at most one- third of the nitrogen needed for the total U.S. corn crop (Fluck and Baird 1980).

However, substitution of manure for chemical fertilizers on corn does result in an energy savings of up to 40% (Lockeretz 1978). Substitution with manures works best when animals and crops are raised in close proximity. The cost of transporting manure more than 20 miles exceeds its fertilizer value. Generally, the use of manures to conserve energy and to improve levels of organic matter is most often feasible on organic farms.

Manure applications are not without problems. Special application ma-chinery is needed, and improper applications can lead to surface runoff carrying nutrients, oxygen-demanding materials, and infectious agents into waterways. Excessive use of manure can cause nitrate pollution of the groundwater, increased soil salinity, and nutrient imbalances leading to poor crop growth. Excellent guidelines do exist for the proper utilization of animal wastes on cropland and pasture land.

In conclusion, manures, which comprise about 22% of the organic wastes in the United States, have been recognized already for their value in reducing nitrogen fertilizer usage, and hence agricultural energy consumption, and in increasing soil organic matter. But, as noted already, an estimated 90% of the annual manure production (175 million dry tons) is already applied on land. A limitation appears to have been reached for this form of energy conservation. Other organic wastes have some nutrient value, but their main use is to maintain organic matter and to control erosion.

### Use of Legumes

Another very old agricultural method for supplying nitrogen, as old as the use of manures, is the utilization of legumes. However, this practice fell into disfavor with the advent of chemical fertilizers. Recently, interest in legumes have been renewed in view of the need to supply nitrogen economically and energy efficiently. Much of this interest centers upon the use of legumes as winter cover crops and in rotations. The amount of nitrogen fixed per acre by various legumes is shown in Table 9.3.

### Table 9.3. Annual Nitrogen Fixation by Legumes

| Crop | Nitrogen fixed[a] (kg/ha) | (lb/acre) |
|---|---|---|
| Perennial | | |
| Alfalfa *(Medicago sativa)* | 193-220 | 172-196 |
| Birdsfoot trefoil *(Lolus corniculalus)* | 58-116 | 52-103 |
| Clover: | | |
| Landino *(Trifolium repens* f. *lodigense)* | 200 | 178 |
| Red *(T. pralense)* | 83-149 | 74-133 |
| White *(T. repens)* | 115 | 102 |
| Kudzu *(Pueraria lobata)* | 120 | 107 |
| Annual | | |
| Bean (edible, *Phaseolus)* | 45 | 40 |
| Lespedeza *(L. cuneata, slriata, stipulacea)* | 95 | 85 |
| Pea | | |
| Cow *(Vigna unguiculata)* | 100 | 89 |
| Garden *(Pisum sativum)* | 80 | 71 |
| Winter *(Lalhyrus hirsulus)* | 60 | 53 |
| Peanut *(Arachis hyogaea)* | 47 | 42 |
| Soybean *(Glycine max)* | 65-110 | 58-98 |
| Vetch (Vicia) | 67-112 | 60-100 |
| Biennial | | |
| Sweetclover *(Melilolus)* | 130 | 116 |

[a] Values taken from numerous sources.

Research indicates that legumes can supply substantial nitrogen to crops that follow. One very promising method is the use of legumes with no-tillage corn. Instead of using the residue of the preceding corn crop to prepare the mulch cover, a legume winter cover crop is used. The benefits of this ap-proach are several, but our interest at this point is the energy conservation achieved through the use of less fertilizer.

Results with this approach have been encouraging. In one study, corn was grown in a no-tillage system with a winter cover crop of hairy vetch. No nitrogen fertilizer was used, but corn yields were equivalent to those observed with 99 kg/ha (87Ib/acre) of nitrogen. With mixed red clover and hairy vetch as the cover crop, corn yields were equivalent to those obtained with 112 kg/ha (100 Ib/acre) of nitrogen. Several other reports cite similar values for hairy vetch. Big flower vetch also seems to give similar results (Frye and Phillips 1981; Frye *et aL. 1983).*

Top yields of corn are possible with legumes supplemented by modest amounts of nitrogen fertilizer. The agronomic value of this practice is clear. Its economic value, while presently borderline in terms of legume seed versus fertilizer costs, will become increasingly positive as prices of fertilizer escalate. In addition use of legume cover crops provide important erosion control benefits (Frye *et aL.* 1983; Palada *et aL. 1983).*

One estimate of the value of legumes in terms of energy conservation has been made by Frye *et aL.* (1983). Legumes, if used on one-half of the no-till corn acreage, would conserve more than 50 billion megajoules of energy or 8.5 million barrels of petroleum annually by the year 2000.

Legumes can also be used in rotations with conventional crops in order to conserve energy. Numerous rotational schemes with corn are possible with annual and perennial legumes, especially those that are useful as forage crops. Some examples follow. Soybeans, as annual legume, can be alternated on a I-year basis with 2 years of corn. This rotation reduces energy needs by 26% compared with continuous corn (Heichel 1978). It must be realized that soybeans, like other annual legumes, do little to reduce the nitrogen needs for the corn that follows. Soybeans receive about 30% of their nitrogen from the atmosphere and about 70% of this is removed with the harvested crop.

Perennial legumes generally have better rates of nitrogen fixation and, hence, the potential to leave nitrogen for crops that follow. These perennial legumes can be used for forage in a rotational scheme with crops such as corn, wheat, and soybeans. The yield of the food crops may be reduced through loss of space to the forage legume, but increasing prices and demand for forage crops as cattle feeds are making such rotations increasingly cost effective (Heichel 1978; Lockeretz 1983). However, an alter-native is to overseed legumes with corn and not lose space to legume rotations. The legume continues growth after harvest through early spring and is then plowed under. A new planting of corn and legumes follows. Early research shows that yields appear similar to straight corn; fertilizer needs and erosion are reduced (Palada *et al. 1983).*

The legume of choice for forage purposes and energy reduction appears to be alfalfa, as it leaves substantial amounts of nitrogen in the soil for succeeding grains. After one harvest, the nitrogen left in the roots and unharvested regrowth of a 2- or 3-year-old stand of alfalfa is equivalent to 190 Ib/acre (213 kg/ha) of nitrogen. Proper use of alfalfa in rotational schemes with food crops may reduce nitrogen needs up to

50% (Heichel 1978). Alfalfa used in rotations with corn can reduce energy input compared with continuous corn by 39%. Rotations with corn, soybeans, wheat, and alfalfa can reduce energy input over continuous corn by 44%. These energy sav-ings are attributed in the largest part to reduced usage of nitrogen fertilizer, but some savings are due to reduced need for fuel and other inputs (Heichel 1978). Triplett *et al.* (1979), based upon their research, suggest that a good meadow stand of vigorous alfalfa can supply nearly all the required nitrogen for a succeeding grain crop, whether no-tillage or conventional tillage is used.

The value of perennial legumes in conserving energy can undoubtedly be im-proved through plant breeding. Some improvements in the nitrogen-fixing ability of alfalfa has been achieved in breeding programs (Heichel *et al.* 1981a; Barnes *et al.* 1981), but much work remains to be done.

Much research also remains in terms of our understanding of legumes. The present methods used to assess nitrogen contributions from legumes appear to be questionable (Heichel and Barnes 1984). For example, they do not account for incom-plete recovery of nitrogen by succeeding crops, nor are they capable of differentiating between normal soil-derived or symbiot-ically fixed nitrogen. It also appears that rotations improve yields for reasons in addition to increased nitrogen; yield increases have been observed in rotations involving only non legumes versus those with legumes and non-legumes. Present methods fail to differentiate between benefits from symbi-otically fixed nitrogen and the other poorly defined cofactors. Still Heichel and Barnes (1984) indicate forage legumes could replace 25-50% of nitro-gen fertilizer needs of high-yield cropping systems.

We also need to better understand the effect of different management practices upon nitrogen contributions from legumes. For example, an unhar-vested legume plowed under would have a greater nitrogen benefit than the residues from harvested forage legumes. However, the amount in the residue that becomes available is a function of the time of soil incorporation, the amount removed for forage, and the frequency of removal (Heichel *et al.* 1981 b). As Heichel and Barnes (1984) have pointed out, the constraints imposed by various management practices upon the substitution of biolog-ically fixed nitrogen for chemical fertilizers need to be clearly expressed.

The problems notwithstanding, the importance of legumes to energy con-servation is significant. As a final perspective, if one-quarter of the nation's corn crop were grown in an annual rotation with alfalfa, the annual savings in nitrogen fertilizer would free up 70 billion cubic feet of natural gas.

**Field Operations: Farm Machinery**

As mentioned already, the energy consumption associated with the use of farm machinery includes direct and indirect components. The indirect component reflects the energy inherent in construction materials and the energy used to manufacture and repair farm machinery. The farmer has little control over this manufacturing energy, and efforts to conserve such energy depend on market forces and the availability and cost of alternative materials and manufacturing processes. However, a program of energy conservation is possible for the direct component of energy consumption by farm machine-ry. Most of this direct energy usage is related to fuel consumed during tillage, planting, pest control, and harvest.

Fuel used in farm machinery represents a significant part of the energy consumed in crop production; operation of farm machinery consumes on an average about 20% of agricultural energy (Lockeretz 1982). For example, fuel accounts for 30% and 42% of the FEF for corn silage and alfalfa, respectively. Decreased or more efficient utilization of farm ma-chinery could be a reasonable avenue for energy conservation. One possible way to decrease fuel usage is through less tillage, which is explored in the next section.

The type of fuel used-gasoline, LP gas, or diesel-has a bearing on energy consumption. In 1960 most farm machinery was fueled by gasoline and LP gas. Today, most new farm machinery is diesel powered. Diesel engines convert fuel more efficiently to mechanical energy, and diesel fuel is both the least expensive and most energy rich per gallon. These advantages far outweigh the slightly higher prices of diesel engines over others. When you consider that fuel and oil to operate farm machinery costs U.S. farmers near $5 billion per year, a powerful economic stimulus exists for fuel conservation.

Energy also can be saved through proper selection, operation, and main-tenance of farm machinery. Equipment most amenable to energy reduction appears to be the tractor and tillage attachments. Comparative test data on the performance of tractors is available from the Department of Agricultural Engineering, University of Nebraska (Lincoln, NE 68503), which operates a tractor-testing facility. Results from tractor tests can be used in a manner similar to that of purchasing a new car. For example, recent tests show considerable differences (23%) in fuel economy at maximal power takeoff. A

number of factors have been shown to affect fuel economy: weight to power ratio, engine efficiency, number of drive wheels, weight distribution, and soil characteristics. Selection of implements for primary tillage is also critical. A moldboard plow produces a clean tilled seedbed, but uses 50% more energy than a chisel plow and 70% more than a heavy offset disk (Bloome *et* al. 1981). If a seedbed with small amounts of crop residue can be tolerated, considerable energy savings are possible. Implement width is also important in tractor selection. The type and width of implements determine what horespower is required at the tractor's drawbar output. Several factors deter-mine implement width: speed of movement in the field, timeliness, and field efficiency. These factors all require careful balancing for most efficient ener-gy usage and economic return.

Field efficiency is expressed as (actual work rate)/(theoretical work rate). Values normally range from 70 to 85%. Time spent on nonproductive opera-tions-turning, making adjustments, overlap between runs, and repairing breakdowns-reduces efficiency below 100%. The higher the speed of tillage and the wider the implement width (up to a point), the greater the actual work rate. This results in increased productivity and reduced labor intensity. The increased horsepower needed per acre to increase productivity and reduce labor does have a cost: The initial investment in equipment is higher. This must be weighed against the economic advantage of increased productivity. Energy savings are therefore possible through downrating of tractor horsepower. However, this must be weighed against timeliness, which accounts for much of the increased yield associated with increases in horse-power. Timeliness is essentially the ability to conduct an activity during such a time as to maximize product quality and quantity. Timeliness is critical at certain periods: planting, harvesting, and sometimes pest control. Yields can be reduced 5-10% if operations are carried out only a few days from the optimal time.

In general, the tractor with the highest fuel economy should be selected. However, it should also have sufficient speed and drawbar horsepower out-put to handle the implement width that allows timely operations at the most critical periods. This would seem to call for overpowered and oversized equipment, but it is justified in terms of increased yield.

It must also be remembered that during those periods when timeliness is not so critical, fuel energy needs are less and the tractor can be run at optimal fuel economy. This fuel usage improvement is possible because diesel engines are most economical when operated at two-thirds of their rated power and rated speed. Tests of this concept, popularly called "gear up and throttle down," have demonstrated average fuel savings of 27% com-pared to full throttle runs (Fluck and Baird 1980; Bloome *et* al. 1981). Care must be taken to avoid overloading the engine.

Future tractors will undoubtedly have small computerized monitors that show energy consumption as affected by numerous variables: throttle set-ting, gear changes, weight distribution (ballasting), tillage depth, and so on. The operator will be able to set conditions to maximize fuel economy for most operations, but to maximize work rate during critical periods.

An important aspect of saving energy and money is proper maintenance of farm machinery. Good maintenance insures that equipment will run effi-ciently, last longer, and go for longer periods of time between repairs. Engine maintenance is of special importance; the best fuel economy comes from a well-tuned engine and clean air filters. Proper maintenance can increase fuel efficiency by up to 25% (Bloome *et al.* 1981).

Other factors can contribute to more efficient usage of fuel. Unnecessary idling should be eliminated: If you take a break, turn it off. Whenever possible, field operations should be combined so as to reduce the number of passes through a field. For example, the formation of a seedbed, fertilization, and sowing of seed might be done in one pass. Tillage reduction, to be discussed next, might be considered. Finally, the use of tractor weight for ballast is critical. Too much weight causes soil compaction and greater use of fuel, while too little means increased tire wear and wheel slippage.

**Field Operations: Tillage**

Tillage, or the working of soil for plowing, planting, and cultivation, is a significant part of agriculture. Estimates are that tillage operations move enough soil in the United States per year to create a superhighway from Los Angeles to New York (Frye and Phillips 1981).

Tillage appears to have peaked in the 1970s. Increased energy costs have encouraged farmers to consider no-tillage and reduced tillage systems. The energy requirements for conventional tillage (moldboard plow), reduced tillage (chisel plow, disk), and no-tillage systems are shown in Table 3.4. Although a no-tillage systems uses only about half the fuel required by conventional tillage, it must be remembered that tillage uses far less energy in crop production than fertilization. Switching from conventional to limited tillage would result in a 3% reduction in the energy requirements for produc-tion of corn (Fluck and Baird 1980); switching to no-tillage corn would give a 7% reduction (Phillips *et al.* 1980). In contrast, the use of legumes on a rotational basis with corn can reduce energy consumption about 10 and four times greater than switching to limited and no-tillage, respectively (Heichel 1978).

However, tillage reduction is being utilized not only to conserve energy but more importantly to reduce soil erosion and, to a lesser degree, increase water filtration and conservation. Since the loss of topsoil is perhaps the most serious problem in agriculture (Sampson 1982), the move to reduced tillage is important. The scarcity of water supplies in some parts of the United States is a further impetus to adoption of reduced tillage.

With reduced tillage (also called minimum or conservation tillage) crop residues are incorporated nearer the surface, since the chisel plow and the disk produce more shallow tillage than the conventional moldboard plow. This acts to increase water filtration and to reduce soil erosion. Reduced tillage requires less mechanical power than conventional (less depth of soil to move), and hence uses less fuel (Table 4).

**Table 9.4. Estimated Energy Requirements for Several Field Operations and Inputs in Four Tillage Systems**

| *Input or operation* | *Gallons diesel fuel/acre* | | | |
|---|---|---|---|---|
| | *Conventional tillage* | *Chisel plow* | *Disk* | *No-tillage* |
| Moldboard plow | 1.84 | | | |
| Chisel plow | | 1.12 | | |
| Disk | 0.63 | 0.63 | 0.63 | |
| Apply herbicides and disk second time | 0.73 | 0.73 | | |
| Spray herbicides | | | | 0.13 |
| Plant | 0.43 | 0.43 | 0.43 | 0.50 |
| Cultivate (once) | 0.42 | 0.42 | 0.42 | |
| Herbicides | 1.75 | 2.01 | 2.25 | 2.88 |
| Machinery and repair | 1.86 | 1.61 | 1.25 | 0.60 |
| *Total* | 7.66 | 6.95 | 5.71 | 4.11 |

*Source:* Frye and Phillips (1981).

No-tillage systems are even more effective than reduced tillage systems for conservation of soil, water, and energy. However, the use of herbicides is necessary to control weeds normally eliminated by cultivation, and increased amounts of other pesticides often are needed to control insects harbored in the large amounts of crop residue. Even counting the energy required to apply pesticides, a no-tillage system is the most energy efficient (Table 9.4). While reduced tillage is acceptable to both conventional and organic farm-ers, no-tillage systems are not acceptable to organic farmers because they require use of pesticides.

Other advantages of reduced and no-tillage have been noted (Fluck and Baird 1980; Frye and Phillips 1981). Higher yields are possible, especially with well-drained land and dryland farming; reductions of labor and time also occur. In some instances double-cropping potential has been advanced northward by several hundred miles as a result of intensification of land use.

Sometimes soil structure is improved. Lower-quality land (e.g., with slopes up to 20%) may be farmed for row crops, and machinery investment may be decreased with reduced or no-tillage systems.

Besides increased usage of pesticides, other disadvantages of reduced and no-tillage systems have been noted (Fluck and Baird 1980; Frye and Phillips 1981). Some soils with fine textures produce lower yields when farm-ed with reduced tillage than when farmed with conventional tillage. More nitrogen fertilizer is needed for increased yields with no-tillage for some crops, such as corn. This increase of nitrogen means a greater potential for water pollution through nitrate leaching. Fortunately, the increased yield of corn more than compensates for the cost and energy value of the additional fertilizer. Increased problems from insects and rodents have been noted. Larger amounts of seed are required.

It would appear that the advantages of no-tillage and reduced tillage out-weigh the disadvantages. At present about 3% of U.S. cropland is farmed with no-tillage; most of the no-tillage crop is corn. Reduced tillage is used for a wide variety of crops on about 25% of U.S. cropland. States that lead in no--tillage are Indiana, Kentucky, Missouri, and Pennsylvania. Leading states for reduced tillage include Illinois, Iowa, Kansas and Nebraska.

**Irrigation**

About 18% of U.S. cropland is irrigated and quite productive; such acreage produces about 27% of our crops (Jensen and Kruse 1981). In the 1980s, irrigation accounted for 13% of the total energy used in agricultural produc-tion (Price 1981). More recent data indicates a drop to 8.3%. Irrigation ranks third as an energy-consuming component of crop produc-tion (transportation is partly associated with livestock operations). However, if only irrigated land is considered, the relative energy input associated with irrigation rises substantially. As energy costs have risen, the rapid expansion of irrigated farmland has slowed because the added costs started to counter-balance the increased productivity of irrigated land.

The amount of energy needed for irrigation is a function of several vari-ables, the most important of which are pumping lift, required water pressure, frequency of irrigation, the type of irrigation system, and the net depth of the applied water. Pumping lift is the vertical distance between the water source and the discharge point over which the water must be pumped. Increasing the water pressure requires additional energy; a water pressure of 670 kilo-pascals (100 Ib/in.$^2$) is equivalent to lifting water 70.4 m (230 ft).

Lower water pressures of 170-345 kilopascals (kPa) are used to operate sprinkler systems with hand-move and side-roll sprinkler laterals. Highest pressures (over 860 kPa) are needed for traveling and stationary hydraulic gun sprinklers. Moderate pressures of 520-690 kPa are used on center pivot spinklers. In surface irrigation systems dependent on gravity flow, less energy is needed than with sprinkler systems that require pumping. Another system with minimal energy requirements is trickle irrigation, where low water pres-sure is needed. Assuming a water lift of 100 m, the annual energy input for these systems would be about 4000 and 8700 kcal/ha, respectively, for surface irrigation (with an irrigation runoff recovery system) and traveling hydraulic gun sprinklers (Batty *et al.* 1975).

A close look at our largest crop, corn, can be used to put energy require-ments into perspective. In the High Plains of Texas 60% of the crop produc-tion energy for corn is attributed to irrigation; this figure becomes 85% with nitrogen fertilizer added. In the Corn Belt, the energy input is about two-fifths of this value, mainly because irrigation is not used. Energy input for irrigation can be even higher. If pumping is required from deep groundwater aquifers, as in Arizona, irrigation can consume 90% of the crop input energy (Jensen and Kruse 1981).

A number of possibilities for conserving irrigation energy (and water) exist (Stetson *et al.* 1975; Heermann *et al.* 1976; Eisenhauer and Fishbach 1977; Ross 1978; White 1978; Jensen and Kruse 1981). Several of these are dis-cussed in the following sections.

### Reduction of Pumping Frequency

Reducing irrigation pumping frequency to the minimum can save energy. For example, although irrigation at certain times has been proven unneces-sary for some crops, based on historical precedent, it is still done. In Idaho sugar beets that are kept well irrigated through August 1, such that the soil has 200 mm (8 in.) of available water, require no additional irrigation until harvest. The last application makes harvest easier. Once grain has reached the soft dough stage, additional irrigation seems to be of little value.

Care must be taken not to eliminate or reduce irrigation during critical stages of development. The critical stage for most grain crops and corn is the flowering and silking stage, respectively. The heading stage on small grains and flowering stage of soybeans is also critical. Irrigation at these times can return the greatest yield per unit of water. For example, average yields over 3 years with unirrigated corn and corn irrigated at the early silking stage dif-fered by 2820-2950 kg/ha (45-47 bu/acre) in Kansas.

### Pump Efficiency

Pumps must be operated at peak efficiencies through proper mainte-nance. New pumps with the greatest efficiency should replace aging, ineffi-cient equipment. An electrical pump's efficiency is that of the pump, whereas the efficiency of pumps powered by natural gas or diesel fuel is a function of the combined efficiency of the engine and the pump. Optimal efficiency appears to be near 75%. Based on pump efficiency tests, one can determine the approximate reduction in energy use that can be achieved by improving efficiency to the maximal limit. The equation for the potential percentage reduction in annual energy use and costs is (1-Ee/Ea) 100, where the existing efficiency is Ee and the attainable efficiency is Ea.

The potential reduction can be used to ascertain the feasibility of steps needed to raise efficiency. Sometimes this is simply a case of adjustment, maintenance, or parts replacement. At worst it may involve replacement of the entire pump. These steps involve variable capital costs, which must be weighed against the potential reduction of energy costs.

### Improved Management

Better management of irrigation schedules can also result in reduction of energy consumption. The best management practice is to employ soil moisture detection devices coupled with computerized scheduling of irriga-tion. One study in Nebraska showed that corn yields were not greatly reduced when pumped irrigation water was cut up to about 50% over the season. In another study use of computerized irrigation schedules for center pivot sprin-klers resulted in a 27% reduction of energy usage, with no effect on corn yields.

Estimates are that energy savings of 30-50% for water pumping are possible through improvements in management practices and equipment efficiency. Adjustments alone on many pumping plants could improve effi-ciency and reduce energy consumption by 13%.

### System Alterations

Other possibilities for reducing irrigation energy requirements include al-teration of application methods or shifting to other procedures such that energy intensiveness is reduced. One alteration is the conversion of a high--pressure sprinkler system to a lower pressure, such that the total dynamic head is reduced (TDH = pumping lift + pressure head + friction). This conversion requries more than changing nozzles; pumping plant modifica-tions are also required (Jensen and Kruse 1981). Such changes are possible on center pivot systems, but the costs versus energy savings savings require careful evaluation. If a high-pressure center pivot system was of such an age as to necessitate replacement, a newer low-pressure center pivot system would be an ideal energy-saving investment.

As discussed earlier surface irrigation systems are more energy efficient than sprinklers, especially if the surface system is well designed. The distribution requires leveled land, channels of closed pipe or lined but open conduits, and a runoff reuse system. The latter is a small reservoir to trap surface runoff, which is returned to the irrigation system. Since the pumping lift is much less for surface water than for groundwater, energy costs are minimal. The energy requirements for pumping surface water are 15-20% lower than those for pumping groundwater.

In some situations sprinklers could be replaced by a surface system. Some surface systems can be made more energy efficient simply by adding a reuse system. Studies in Nebraska indicate that the addition of reuse systems can reduce pumping costs by 25%, since the recycled runoff reduces the time that the groundwater pumps are operated (Jensen and Kruse 1981). Another more energy-efficient replacement for sprinklers is a trickle system.

The use of gravity pressure rather than pumping lift can cut energy con-sumption. This method is possible by diversion of a mountain stream to a pipeline leading to sprinklers in valleys. Gravity flow of water for surface irrigation as opposed to pumping can also reduce costs. Upstream river diversion at points higher than the surface-irrigated area is required for this approach.

Improvement of irrigation efficiency can also cut energy consumption. Techniques include more uniform application of water, nighttime applica-tions to reduce evaporation, application of only the amount of water retaina-ble in the root zone, reduction of surface runoff, and elimination of leaks and seepage. A most-promising energy-saving innovation is the photovoltaic—powered water pump. Tax credits are available for their installation. Functional units are now in use in the United States (Arco 1982) in areas far removed from utility lines.

**Pesticides**

In terms of the crop production energy budget, pesticides are not nearly as demanding as fertilizers. Fertilizers compose 98% of farm chemicals; pesticides only 2%. For example, pesticides utilized in corn production con-sume about 0.8-3.5% of the energy input (Pimentel *et at.* 1973; Heichel 1978). The production and application of pesticides, according to the Coun-cil for Agricultural Science and Technology (CAST 1977), consumes about one-seventh as much energy as fertilizers. Thus, the elimination of pesticides would not result in a significant conservation of agricultural energy.

However, despite only modest potential savings in energy, there appear to be many compelling reasons for judicious use of pesticides and integrated pest management. Problems with pesticides include development of re-sistance among pests exposed continuously to pesticides and deterioration of environmental quality.

A number of practices can be used to control pests without pesticide use. These include weed control through tillage, mulches, sanitation, closer crop spacing, crop rotation, and proper timing of seeding and planting. A large number of natural controls are possible for suppression of insects and dis-ease.

**Crop Driying**

Annual energy consumption for the drying of crops is at least 0.111 x $10^{12}$ megajoules (Fluck and Baird 1980). The practice of removing moisture from crops (e.g., corn, soybeans, rice, peanuts, and tobacco) with hot air derived from burning fuels, mostly natural or LP gas, has increasingly replaced natural air drying. Crop drying accounts for 3.4% of the total energy consumed during crop production. Dried crops are easier to store because they are less likely to spoil, overheat, or become infested with insects.

At present about half the U.S. corn crop is dried directly on farms and another quarter off the farm. Drying, besides aiding good storage, permits earlier harvesting, the growing of higher-yielding, later-maturing cultivars, and, in some areas, double cropping. These advantages incur an energy debt. For example, the fuel required to dry 1 ton of corn from 25-15% moisture is 18.4 liters of LP gas. Expressed in other terms, the same 10% reduction in moisture for the average yield of corn in a hectare (2.47 acres) would require 156-196 liters (39 to 49 gal.) of LP gas. Some alternative would appear desirable, especially in view of the gas shortages in the Midwest during the recent decade, which curtailed the drying of corn.

Some energy-saving alternatives do exist (Frye and Phillips 1981). These include field drying, low-temperature drying, solar drying, dryeration, and direct preservation of high-moisture grain destined for cattle or hog con-sumption.

One of the simplest and most promising techniques, if the advantages of crop drying have high priority, is dryeration. Grain is removed hot from the drier just short of acceptable moisture contents and slowly aerated-cooled after being held for 4-12 hr with aeration. It may also be possible to dry grain at lower temperatures and slower rates in certain regions. This procedure is feasible for harvests in cool and not overly humid regions, but not in areas like the southeastern United States. In these hotter, more humid areas, it is more feasible to dry first at conventional high temperatures until the moisture is reduced enough to allow storage lengths acceptable for a second drying stage, such as prolonged low-temperature drying or natural air drying.

Delayed harvest results in field drying. However, the savings in energy may be negated by possible yield decreases and additional field losses. Present late-maturing cultivars are often not suitable for delayed harvest, and earlier cultivars frequently yield less. Damp, rainy weather may pose problems in terms of disease and drying. In some areas the potential of a double crop might be lost. One possibility, if planting conditions permit, is earlier planting of later-maturing cultivars, such that harvest occurs at the usual time but with drier grain (Morey *et aL.* 1978). Field drying requires careful assessment to insure that yield losses are not equal to or greater than energy savings.

Solar energy appears to be promising for grain drying, and current tax advantages for solar equipment help to offset the high initial expense. Because grain drying is done only during a short period of time, solar systems must be inexpensive or serve multiple purposes (Spillman 1981). The future looks bright for slow low-temperature solar drying, but not fast high-temperature drying (Chau and Baird 1978). Another promising technology is heat pump driers (Peart *et al. 1980).*

**Frost and Cold Protection**

Protection of frost- and chilling-sensitive crops consumes energy, usually in the form of burning fuel and movement of air by wind machines (Fig. 3.4). The annual energy expenditure is about equivalent to that for greenhouse heating, 0.045 x $10^{12}$ megajoules (Fluck and Baird 1980). Frost and cold protection ranks far below crop drying as an agricultural energy consumer, but its low rank can be deceptive. For example, in some years citrus groves require no frost protection, whereas in other years energy needs for such protection may be 60.8% of the required energy in citrus groves (USDA/FEA 1977). Energy consumption for a typical hectare in a citrus orchard might range from 400 to 1600 liters (105.6 to 422.4 gal.)/hr of oil depending on ambient temperature. The benefits of energy conservation are obvious in these conditions.

Conservation techniques include an assessment of the energy costs for producing the crop versus the energy costs for frost protection. If the weather abnormality persists, it could be economic disaster to continue protection. At some point saving the crop becomes unjustified, as energy costs for produc-ing a new crop become so much less than the energy expended for cold protection. Other possibilities are the use of sprinklers for minor cold protec-tion instead of heaters and wind machines. For example, annual energy input for frost protection of cherries with oil heaters is 84,708,760 kcal/ha but is only 8,161,000 kcal/ha with irrigation (Pimentel 1980). Sprinklers can also be used to delay blossoming through evaporative cooling, thus bypassing unexpected late frosts.

A long-range solution may be found in existing research. Under some conditions water may be supercooled in liquid form to – 5°C; under these conditions, frost protection could be withheld more so than is the acceptable practice. Since most frost protection efforts are attempted in the temperature range of 0° to – 5°C range, much of the energy need for cold protection could be eliminated. However, supercooling is prevented if nucleation cen-ters for the formation of ice crystals are present. Nucleation centers are provided by bacteria associated with soil and plants, such as *Erwina her-bieola* and *Pseudomonas syringae.* Plants themselves have no such nuclea-tion centers. Scientists have isolated a bacterial gene that is involved with the nucleation center (Anon. 1982a). Eventual research may lead to an inhibitor or alterations through genetic engineering, thus reducing energy costs for cold protection. One such inhibitor, a polymer, appears close to reality (Anon. 1982b). Kocid 101, a fungicide and bactericide, became licensed for light frost protection through control of ice-nucleating bacteria in 1983.

**Transportation**

Transportation ranks third as an energy consumer in agricultural produc-tion. Its rank in crop production is somewhat difficult to determine, since transportation also utilizes considerable energy in livestock production. Based on the available data, it is estimated that roughly one-fourth of the transportation activity is associated with animal production. This estimate would mean that transportation associated with crop production uses about the same, or slightly more, energy than irrigation. The relative rank (whether third or fourth) of transportation and irrigation in energy consumption in crop production cannot be accurately determined with the present data.

Attempts to cut energy usage in farm transportation depend presently upon conservation. Vehicles must be properly tuned and lubricated if optimal fuel economy is to be realized. Transportation patterns should be as efficient as possible. Two or three trips with light loads should be replaced with one trip. Several errands over a week might possibly be combined into one trip. As aging vehicles are replaced, newer ones should be as fuel efficient as possible.

In the long run, farm transportation can only become more sustainable if alternate fuels from renewable resources become available. For example, ethanol can be produced through fermentation of grains, such as corn. Ethanol can be mixed with gasoline (gasohol) or used alone to run gasoline--dependent vehicles. Some adjustments are required, but they are possible. A few farmers, how many is unknown, run pickup trucks and cars on ethanol now. Ethanol can also partly supplement diesel fuel, but considerable modifi-cations are involved.

However, widespread adoption of alternate fuels for gasoline and diesel is in the future. A discussion of alternate fuels including present technology, needed improvements, and constraints is presented.

**Future Directions**

The present approach to reducing energy usage in crop production focuses on conservation. As long as energy supplies are sufficient and con-servation measures compensate for price increases, this would seem to be a prudent choice. However, as energy supplies inevitably dwindle in the future, agriculture must eventually move toward energy sustainability. The most logical approach to achieving this is use of alternate energy sources derived from renewable resources.

The prospects of reaching that goal are highly mixed. For example, the biggest energy consumer, nitrogen fertilizer, could become sustainable first through the use of legumes and then with crops having genetically induced biological nitrogen fixation. The use of legumes is presently feasible, and indeed was once the common practice. With more research, the widespread adoption of this practice is not insurmountable if the need arises for an interim or partial solution. The ideal way to reduce usage of nitrogen fertilizer is the genetic engineering of our present crops to meet their nitrogen needs like legumes, that is, through biological nitrogen fixation. Here the prospect is far less promising, until considerable research is accomplished. The major breakthrough needed to realize this goal is not imminent, but more likely far in the future.

The impetus for achieving nitrogen sustainability is the present energy intensiveness of chemical fertilizers plus the availability of a simple renewable resource, atmospheric nitrogen, that could replace them. The push for sus-tainability in terms of phosphate and potash is not as hard, since energy consumption and soil needs for these two nutrients are considerably less. However, the ore resource for phosphate is now seen as limited. Sus-tainability of these nutrients probably could be realized through recycling efforts, for example, by applications of unprocessed or composted organic materials. However, it is not now clear what extent nutrient needs could be satisfied through this approach, nor what would constitute optimal management practices.

Direct use of energy on the farm can also become a sustainable practice. Fuel use accounts for most of the energy consumption associated with field operations and transportation. Farmers presently use some 6 billion gal. of fuel annually, about equally divided between gasoline and diesel. A significant impact upon energy sustainability would result if these were replaced with alternate renewable fuels. The basic technology for doing so is presently available, although improvements and more incentives are needed if wide-spread adoption is to take place. A careful analysis is also required to assess the effects of producing alternate fuels in substantial quantities on food prices, food supplies, and soil erosion. Some improvement in the com-patibility of combustion engines to alternate fuels is also required.

The alternate fuel that presently shows the most promise is alcohol, es-pecially ethanol. The technology for producing ethanol is available and al-ready being used, although to a very limited extent. The fermentation process that produces ethanol also produces animal feed as a byproduct. Ethanol can be used to supplement or replace gasoline with minimal difficulty; it also can supplement diesel fuel although this requires substantial engine modifi-cation. Alternately, diesel tractors could be replaced by tractors with different engines that utilize straight ethanol. Such units exist in Europe. A diesel fuel equivalent can also be prepared from vegetable oils, but substantial research is needed to overcome problems associated with this technology. An in-depth discussion of alternate fuels is presented in on future technology.

Another alternate energy source is solar energy. The basic technology is available, although refinements and some new technology are still needed. The most likely scenario is partial, gradual adoption of solar energy in the near to moderate future, especially for grain drying, hot water heating, and space heating. Solar devices for these purposes have already been adopted by a few users for additional information on solar energy.

The adoption of the preceding approaches to sustainability will be mixed. Which practice or practices and to what extent each will be utilized depends upon not only national needs, but regional and even individual farm charac-teristics. The adoption of sustainable practices and technologies will be influ-enced by several factors: farm size, crop mix, livestock presence, conven-tional energy availability and price, cost of alternate systems, management skills, do-it-yourself ability, and information availability. The future pattern of sustainable crop production will not be uniform, but a mosaic consisting of mixed practices that best meet local needs.

## REFERENCES

Anon, 1982a. Gene may protect crops from frost damage. Chem. [, Eng. News (Aug. 2), 6. ANON. *1982b*. Polymers may be able to block frost in crops. Chem. [, Eng. News (Oct. 4), 31-32.

Anon, 1983. IVA set sights on future phosphate technology. Chem. [, Eng. News (Nov. 7), 37-38.

Arco, 1982. Photovoltaic-powered water pumps growing in popularity. ARCO Solar News 2 (4) 4-9.

Barker, A. V. 1980. Efficient use of nitrogen on cropland in the Northeast. Conn. Agric. Exp. Stn. Bull. 792. New Haven, CT.

Barker, A. V. and MillS, H. A. 1980. Ammonium and nitrate nutrition of horticultural crops. Hort. Rev. 2, 395-423.

Barners,, D. K., Heichel, G. H., Vance, C. P., Viands, D. R. and Hardarson, G. 1981. Successes and failures while breeding for enhanced nitrogen fixation in alfalfa. *In* Genetic Engineering of Symbiotic Nitrogen Fixation. J. M. lyons, R. C. Valentine, D. A. Philips, D. W. Rains and R. C. Huffaker (Editors). Plenum Publishing, New York.

Batty, J., Hamad, S. N. and Keiier, J. 1975. Energy Inputs to Irrigation. J. In. and Drainage Div., ASCE 101 (IR4),293-307.

Bioome, P. D., Grevis-James,1. W., Jones, l. K. and Batchelder, D. G. 1981. Farm machinery ideas that save energy. *In* Cutting Energy Costs (The 1980 Yearbook of Agri-culture). J. Hayes (Editor). U.S. Dept. Agric., Washington, DC.

Cast, 1977. Energy Use in Agriculture. Rept. No. 68. Council for Agric. Sci. and Technol., Dept. of Agronomy, Iowa State, Ames.

Chau, K. V. and Baird, C. D. 1978. Solar grain drying under hot and humid conditions. *In* Proc. 1978 Solar Grain Drying Conf., Purdue University, West lafayette, IN.

Eisenhauer, D. E. and Fischbach, P. E. 1977. Comparing costs of conventional and improved irrigation systems. Irrig. Age 11 (8) 36-37.

Fluck, R. C. and Baird, C. D. 1980. Agricultural Energetics. AVI Publishing Co., Westport, CT. FOX, R. H. and PIEKIEIEK, W. P. 1978a. A rapid method for estimating the nitrogen-supply-ing capability of soil. Soil Sci. Soc. Amer. J. 42, 751-753.

Fox, R. H. and Piekieiek, W. P. 1978b. Field testing of several nitrogen availability indexes, Soil Sci. Soc. Amer. J. 42, 747-750.

Frye, W. W. and Phillips, S. H. 1981. How to grow crops with less energy. *In* Cutting Energy Costs (The 1980 Yearbook of Agriculture). J. Hayes (Editor). U.S. Dept. Agric., Washington, DC.

Frye, W. W., Herbek, J. H. and Blevins, R. L. 1983. legume cover crops in production of no-tillage corn. *In* Environmentally Sound Agriculture. W. lockeretz (Editor). Praeger Pub-lishers, New York.

Heermann,, D. F., Haise, H. R. and Mickelson, R. H. 1976. Scheduling center pivot sprinkler irrigation systems for corn production in eastern Colorado. Trans. ASAE 19, 284-287.

Heichel, G. H. 1977. Energy analyses of forage production. *In* Proc. Intern. Conf. on Ener-gy Use Management, Vol. 1. R. A. Fazzolare and C. B. Smith (Editors). Pergamon Press, New York.

Heichel, G. H. 1978. Stabilizing agricultural needs: Role of forages, rotations, and nitrogen fixation. J. Soil and Water Cons. 33, 279-282.

Heichel, G. H. and Barnses, D. K. 1984. Opportunities for meeting crop nitrogen needs from symbiotic ntirogen fixation. *In* Organic Farming: Current Technology and Its Role in a Sustainable Agriculture. J. Power and D. Bezdicek (Editors). Amer. Soc. Agro., Madison, WI.

Heichel, G. H., Vance, C. P. and Barnes, D. K. 1981a. Evaluating elite alfalfa lines for NT fIXation under field conditions. *In* Genetic Engineering of Symbiotic Nitrogen Fixation and Conser-vation of Fixed Nitrogen. J. M. lyons, R. C. Valentine, D. A. Phillips, D. W. Rains, and R. C. Huffaker (Editors). Plenum Publishing Corp., New York.

Heichel, G. H., Barnes, K. D. and Vance, C. P. 1981 b. Nitrogen fixation by forage legumes, and benefits to the cropping system. *In* Proc. 6th Annu. Symp. Minnesota Forage and Grassland Council. St. Paul, MN.

Jensen, M. E. and Kruse, E. G. 1981. Cheaper ways to move irrigation water. *In* Cutting Energy Costs (The 1980 Yearbook of Agriculture). J. Hayes (Editor). U.S. Dept. Agric., Washington, DC.

K;p[atel, J. M. 1978. Energetics of managed land-use systems. *In* Proc. Intern. Conf. on Energy Use Management, Vol. 1. R. A. Fazzolare and C. B. Smith (Editors). Pergamon Press, New York.

Lobehardt, W. C. 1983. Variability of fertilizer recommendations by soil testing laboratories in the Unites States. *In* Environmentally Sound Agriculture. W. Lockeretz (Editor). Praeger Pub-lishers, New York.

Lpcleretz, W. 1978. Economic and resource comparison of field crop production on organic farms and farms using conventional fertilization and pest control methods in the midwestern United States. *In* Towards a Sustainable Agriculture. Jean-Marc Besson and Hardy Vogtmann (Editors). International Federation Organic Agriculture Movement Con-ference 1977. Verlag Wirz AG, Aarau, Switzerland.

Lpcleretz, W. 1982. Energy use by agriculture: How big a problem? *In* Proc. Cornucopia Project Symposium: Planning a Sustainable U.S. Food System, Lehigh Univ., 1981. Rodale Press, Emmaus, PA.

Lpcleretz, W. 1983. Energy in U.S. agricultural production. *In* Sustainable Food Sys-tems. Dietrich Knorr (Editor). AVi Publishing Co., Westport, CT.

Morey, R. V., Gustafson, R. J., Cloud, H. A. and Water, K. L. 1978. Energy require-ments for high-low temperature drying. Trans. Asae 2], 562-567.

Olson, R. A., Frank, K. D., Grabouski, P. H., Rehm, G. W. and Knudsen, D. 1982. Economic and agronomic impacts of varied philosophies of soil testing. Agron. J. 74, 492-499.

Palada, M.C., Ganser, S., Hofstetter, R., Volak, B. and Culik, M. 1983. Association of interseeded legume cover crops and annual row crops in year-round cropping systems. *In* Environmentally Sound Agriculture. W. Lockeretz (Editor). Praeger Publishers, New York.

Peart, R. M., Brook, R. and Okos, M. R. 1980. Energy requirements for various methods of crop drying. *In* Handbook of Energy Utilization in Agriculture. D. Pimentel (Editor). CRC Press, Boca Raton, FL.

Phillips, R. E., Blevins, R. L., Thomas, G. W., FRYE, W. W., and Phillips, S. H. 1980. No-tillage agriculture. Science 208, 1108-1113.

Pimentel, D. (Editor) 1980. Handbook of Energy Utilization in Agriculture. CRC Press, Boca Raton, FL.

Pimentel, D., Hurd, L. E., Bell Toti, A. C., Forster, M. J., Oka, LN., Sholes, O. D., and Whitman, R. J. 1973. Food production and the energy crisis. Science *182, 443-449.*

Price, D. R. 1981. Where farm energy goes. *In* Cutting Energy Costs (The 1980 Yearbook of Agriculture). J. Hayes (Editor). U.S. Dept. Agric., Washington, DC.

Ross, R. 1978. Colorado pump tests show how to make big dollars savings. Irrig. Age (March), 12(3) 9, 12, 16.

Sampson, N. 1982. Soil, land use and abuse. *In* Proc. Cornucopia Project Symposium: Planning a Sustainable U.S. Food System, Lehigh Univ., 1981. Rodale Press, Emmaus, PA.

Spillan, C. K. 1981. Capturing and storing energy from the sun. *In* Cutting Energy Costs (The 1980 Yearbook of Agriculture). J. Hayes (Editor). U.S. Dept. Agric., Washington, DC.

Stetson, L. E., Watts, D. G., Corey, F. C. and Nelson, I. D. 1975. Irrigation system management for reducing peak electrical demands. Trans. ASAE ] 8, 303-306, 311.

Triplett, G. B. JR., Haghiri, F., Van Doren, D. M. 1979. Legumes supply nitrogen for no-tillage corn.

Ohio Report (Nov.-Dec.), 83-85.

USDA. 1982. Agricultural Statistics 1981. U.S. Dept. Agric., Washington, DC.

USDA/FEA. 1977. A Guide to Energy Savings for the Orchard Grower. U.S. Dept. Agric,, Washington, DC.

USDA/FEA. 1976. Energy and U.S. Agriculture, 1974 Data Base. U.S. Dept. Agric. and Fed.

Energy Adm., Washington, DC.

Vilstrup, D. 1981. Less energy, more food. *In* Cutting Energy Costs (The 1980 Yearbook of Agriculture). J. Hayes (Editor). U.S. Dept. Agric., Washington, DC.

Waldrop, M. M. 1981. Deep changes taking root in U.S. agriculture. Chem. f., Eng. News (June 1), 23-28.

White, J. G. 1978. Re-use pits: cheapest water on the farm. Irrig. Age (Nov.-Dec.), 12(11), 58,62.

# 10

# Agriculture's Impact on the Environment

Agricultural activities have had a definite impact upon the surrounding environment. Effects that have been noted include water, soil, and air pollu-tion. In turn, indirect effects upon animal and even human life have been observed. Unhappily these impacts have resulted in a reduction of environ-mental quality.

Water quality has declined as a result of agricultural pollutants, such as sediment, nutrients, salts, animal wastes, and pesticides. Air quality has suf-fered from airborne particulates produced by wind erosion. Pesticides have been harmful to fish and wildlife and have posed a threat to humans when the chemicals become incorporated into the food chain. Animal wastes have contributed to microbial pollution of water, thus posing a disease threat to animal and human life.

An examination of these problems follows, after which agricultural prac-tices that sustain the environment will be covered.

***Water Quality***

Pollutants may enter water from a specific defined location (point source) or as a result of runoff from land (nonpoint source). The major cause of nonpoint source pollution is agriculture. Storm runoff, because of present agricultural practices, carries pesticides, particles of soil, nutrients, and organic wastes. Irrigation water return is also a non point source pollutant, when it is saline and contains nutrients.

Sediment forms the largest agricultural pollutant. In fact, the number one pollutant of water by volume is sediment. Erosion of cropland contributes about half of the total sediment, somewhere around 760 million tons a year (USDA 1981). Forestry, construction, and mining contribute the rest.

The detrimental effects of sediment are several fold. It fills in lakes, reser-voirs, and rivers, causing displacement of water, which can hinder flood control efforts, increase flood hazards, and even end the usefulness of reser-voirs. Eventually, the recreational value and electric power production capac-ity of bodies of water can be diminished. Costly cleaning and dredging may be necessary.

Sediment pollution also reduces populations of fish and shellfish, and decreases the capacity of waters to assimilate oxygen-demanding wastes. This is caused in part by suspended solids cuttling sunlight, thus reducing the growth of aquatic plants and subsequently those animals that feed on plants. Settled solids form an oxygen-demanding sludge. In addition, as discussed in Chapter 8, sediments carry substantial amounts of nitrogen, phosphorus, and low concentrations of certain pesticides. Nutrients lead to eutrophication and sometimes hazardous levels of nitrates. Pesticides can cause fish and animal deaths, and some can be hazardous to humans if they move through the food chain.

The aqueous portion of runoff carries some soluble salts, pesticides, and even microorganisms. Recycled irrigation water, when returned, can raise salinity levels. Soluble nutrients lead to eutrophication; that is, they stimulate growth of algae and water weeds, ultimately causing a reduction of dissolved oxygen. Fish populations and recreational quality decline. Salinity, if high enough, leads to crop damage, unpotable water, and even elimination of some aquatic life. Microorganisms and soluble salts can enter runoff arising from feedlots and sites of manure storage. Some microbes from wastes may be hazardous to humans.

Methods for assessing the deterioration of water quality by these various pollutants are described in an EPA manual (EPA 1982).

**Air Quality**

Agricultural activities do not lead to serious air pollution. In fact, the two primary air pollutants, oxides of sulfur and nitrogen, arise from industries and utilities that burn coal, oil, or natural gas. These gaseous pollutants can undergo further reactions to produce secondary pollutants. For example, the oxides of nitrogen and sulfur can be oxidized further, and when combined with precipitation, produce the well-known acid rain. Photochemical conver-sion of nitrogen oxides and unburned hydrocarbons (from internal combus-tion engines and industry) leads to ozone formation. Acid rain appears to pose no present threat to food crop production, but accumulating evidence suggests a potentially harmful impact upon forest crops (Klein and Klein 1985). Gaseous pollutants, however, are clearly a threat to food production.

Certain types of air pollution pose a considerable threat to agriculture (Kress and Miller 1983; Miller 1983). Sulfur dioxide (S02) in low concentra-tions improves soybean productivity, especially in soils where sulfur availabili-ty is limited. Essentially, the sulfur dioxide serves as a nutrient source. Howev-er, as S02 exposure approaches 10 ppm-hours, yields decrease. Decreases near 20% have occurred as the S02 dose nears 20 ppm-hours (Miller 1983).

On the other hand, nitrogen dioxide ($N0_2$) has little or no effect on soy-bean yields. However, when combined with S02' a powerful synergistic effect occurs. Concentrations of $N0_2$ that are harmless alone have caused yield losses of soybeans of 20% and even more when mixed with marginally harmful S02' Low levels of each (0.13 ppm $N0_2$ and 0.17 ppm S02) when combined reduced seed yields by 18% (Miller 1983).

Ozone also damages plants. Soybeans are highly sensitive (Kress and Miller 1983), and corn, sorghum, and winter wheat are also susceptible, but less sensitive than soybeans. Acid rain appears to pose no threat to these and other crops (Miller 1983).

The main air pollutants arising from agricultural activities are soil particu-lates and pesticides. Some 30 million tons of particulates are placed into the air annually because of wind erosion of soil. The seriousness of uncontrolled particulate pollution can be vividly demonstrated during the Dust Bowl era of the 1930s. The health of animals and even people were affected by the dust storms. Today, particulates are not nearly as serious, but they do contribute to increased maintenance of machinery, lung and throat irritation, aggrava-tion of respiratory diseases, reduced visibility, and deterioration of buildings. They may even reduce agricultural productivity by decreasing the quality of light reaching plants and by partially obstructing stomates.

Pesticides can also appear in the air through application activities. While air-borne, pesticide droplets or particulates can be hazardous if they are inha-led. Once settled from the air, pesticides can be hazardous to fish and animal life, or even to humans, if the chemicals become part of the food chain.

## Soil Quality

The most troublesome agricultural pollutants in terms of soil quality are pesticides. Certain toxic wastes, if used as fertilizers or organic amendments, can also be harmful. Both can have an effect upon soil invertebrates and micro-organisms.

Responses of soil invertebrates to pesticides are complex, variable, and not thoroughly researched. About the only possible generalization is that pesti-cide use does change the function and structure of soil communities. Direct and indirect responses by soil invertebrates to pesticides are found. Popula-tions may decrease, but some increase. Some species are destroyed, yet pesticides can accumulate in some organisms without apparent harm. The effects of individual pesticides upon the soil community are understood somewhat, but the effect of multiple pesticides upon the soil community needs much more research.

Pesticides have been known to affect soil microorganisms. This is ob-served with certain pesticides, but especially with fungicides and fumigants; effects may persist for a long time. The seriousness of microbial changes is evident when microorganisms causing plant diseases or regulating nutrient cycles increase and decrease, respectively.

An area in need of research is whether the use of pesticides damages the soil ecosystem to the extent that crop productivity is lessened over a long period of time. Present data are sparse. Many insecticides and herbicides have the potential to suppress or destroy soil microorganisms if their con-centrations become excessive. However, with normal application rates and reasonable mixing of soil through tillage, these chemicals usually fail to reach soil concentrations above 3 ppm. The lack of soil mixing and heavy herbicide usage with no-till and minimal till systems suggests a potential problem needing research. Even if their persistence is limited, the toxicity of degrada-tion products is poorly understood. Ultimately, soil microflora and even crops might be affected, either indirectly through alteration of the soil ecosystem or directly through phytotoxicity.

**Pesticides of particular concern include fungicides and soil fumigants.**

While targeted at pathogens, they usually have a broad spectrum and destroy beneficial microflora. Thus, part of the soil ecosystem is temporarily harmed. Again the long-term effect of residuals is poorly understood.

The present consensus by soil microbiologists is that the short-term ef-fects of pesticides do not warrant their ban. The one short-term problem may be with no-till agriculture, because of extensive herbicide use. Clearly, more research is needed on extensive herbicide use and the long-term effects of pesticides (OTA 1982).

Some forms of organic wastes, when used as fertilizers, have the potential to introduce toxic substances into cropland. These toxins include heavy metals and polychlorinated biphenyls, which are most likely to be present in sewage sludge and organic industrial wastes. This issue is discussed further by Parr *et al. (1983)*.

Heavy metals do threaten soil organisms and humans if they become part of the food chain. Soil microbial processes, especially ones of agricultural value like microorganismal decomposition of organic matter and the nitro-gen cycle, are inhibited by heavy metals. The level of inhibition is dependent upon the type of heavy metal, its concentration, and the soil type and pH (OTA 1982).

## Reducing Agricultural Pollutants

The major share of pollution resulting from agricultural activities can be eliminated. The main pollutant, sediment, arises from erosion. Prevention or control of erosion would accomplish two purposes: sustain ability of an agri-cultural resource and reduction of pollution. The need to control erosion is compelling if agriculture is to be sustained with minimal environmental threat. Practices of erosion control were thoroughly discussed in further chapters.

Judicious use and adoption of certain approaches when using organic wastes, fertilizers, irrigation water, and pesticides could minimize the harmful effects of chemicals, nutrients, salts, and pathogens upon the environment. Ways to minimize salinity when using irrigation water have been covered in Chapter 8. Proper handling of manures and other wastes in terms of minimal pollution. Since most of the nutrients ending up as water pollutants are carried by sediment (Bradford 1974; Alberts *et al.* 1978), the most effective practice to minimize nutrient pollution is to control erosion.

One effective technique not mentioned previously is to use fertilizers judi-ciously. This has a double benefit: It minimizes pollution and conserves energy. Excess use of fertilizer is not only energy wasteful, but costly and likely to increase pollution. Some additional research is needed to establish ideal quantities of fertilizer based upon crop and soil types. This information must be communicated to the farmer along with the need for frequent soil testing. If farmers realize that money can be saved by judicious use of fertil-izers, they are likely to adopt this approach. The secondary impetus would be to promote the fact that the farmer is sustaining agriculture by using less energy and improving the environment.

In the remainder of this chapter, the primary approach to reducing the environmental impact of pesticides is discussed.

## Integrated Pest Management

The most reasonable way to minimize pollution by pesticides and yet still retain control over pests is to use integrated pest management (IPM). Interest in and adoption of IPM is such that IPM is becoming a mainstay of sustainable agriculture.

The concept of IPM has been defined by the United Nations Food and Agriculture Organization in the following manner: "A pest management sys-tem that, in the context of the associated environment and the population dynamics of the pest species, utilizes all suitable techniques and methods in as compatible a manner as possible and maintains the pest populations at levels below those causing economic injury." Such a management system uses several approaches to pest control. Chemical control with pesticides remains, but as just one part of a coordinated control strategy that also encompasses biological, cultural, competitive, and biorational controls. These components have been well defined by Batra (1982). Two good over-views on IPM are those of Allen (1980) and Frank (1981).

**Resource limitation is one of several factors fostering adoption of IPM.**

Pesticides, being petrochemical products, will become increasingly expen-sive in terms of both money and energy consumption in the agricultural budget. With its judicious and sparing use of pesticides, IPM appeals to practical-minded farmers and those with concerns for future generations of farmers. Adoption of IPM practices has resulted in real savings, as discussed later in the section on economic benefits.

**Environmental concern is another compelling reason for adoption of IPM.**

Pimentel and Edwards (1982) present a good summary of the well-known problems with pesticides: Some kill beneficial insects or cause other disrup-tions of natural ecological cycles, and because of their persistence in the environment, some pesticides can kill fish and wildlife and ultimately con-taminate the groundwater and organisms in food chains.

One example of a pesticide appearing in the food chain is ethylene dibromide (EDB), which has been shown to be a serious carcinogen in animal tests (Hanson 1984). One use of EDB was for the fumigation of stored grains and milling machinery, and residues of it appeared in some flour, cake mixes, and other grain products. Another use of EDB was as a soil fumigant; as a result it eventually appeared in the ground-water in Georgia and, more seriously, in Florida. Fruit and vegetable contamination also arose, since EDB was used as a postharvest fumigant. Ultimately, the prevalence of EDB contamination caused the Environmental Protection Agency to halt agri-cultural uses of EDB. Ironically, EDB had replaced a previous soil fumigant, dibromochloropropane, which had been banned a few years before.

Besides environmental and health concerns, another problem is the devel-opment of resistance to pesticides. As a result, pesticides become in-creasingly ineffective, leading to pest control failures. An example is the failure of pesticides to control cotton pests in northeastern Mexico and south-ern Texas in the late 1960s and early 1970s (Adkisson *et al. 1982).*

A more recent example of pesticide failure is the inactivation of certain pesticides by soil microorganisms (Fox 1983). Microbes in some soils now degrade pesticides so quickly that the efficacy of the pesticide is rapidly diminished. This effect has been noted with several herbicides and some insecticides. Affected crops include corn in the midwestern United States, vegetables, and managed forests. The extent and understanding of this phe-nomenon is poorly known, since it was only recently discovered.

Other factors that argue for the IPM approach over use of pesticides alone are the decreased crop yields and increased diseases associated with some pesticides (Friend 1983; Hodges and Scofield 1983). For example, yields of lettuce have been decreased by several common pesticides as a result of their adverse effects upon photosynthesis and transpiration. The use of some herbicides has been associated with increased susceptibility to insects and disease. For example, the use of 2,4-D on corn increased the susceptibility to southern corn leaf blight *(Helminthosporium maydis)*, corn leaf aphid *(Rhopalosiphum maidis)*, and the European corn borer *(Ostrinia nubilalis)*.

In spite of widescale pesticide usage, it is estimated that pests cause losses of 33% of annual potential production, worth $20 billion (Batra 1982). These losses lead to an ever-escalating production of new pesticides and increasing costs. Insecticide use has increased 10-fold since the 1940s, yet crop losses from insects have doubled over this same period (Pimentel *et al.* 1983). In contrast the stability and minimal resistance associated with IPM makes it quite attractive.

An economic analysis of corn pest management for the 1980 crop (Hant-horn and Duffy 1983) suggests that pesticides are cost effective, but just barely. For example, farmers who used herbicides realized an estimated return of $1.05 for each $1 spent on herbicides; the return for $1 spent on insecticides was $1.03. Since corn is a major crop, the use of pesticides producing only marginal returns has important implications. However, addi-tional studies are needed to verify this finding. A complication is that 1980 was a drought year. Moisture stress has an effect on yield and the effect of such stress on pesticide effectiveness is difficult to assess. Secondly, pesti-cide costs were based on national averages, as were returns per bushel of corn. This assumption for pesticide costs may have overstated actual prices paid by farmers. Still, studies like these make IPM look increasingly attractive.

Technological advances with pesticides have helped to make IPM more feasible. While sounding contradictory, two advances have helped to allow better combinations of pesticides with non chemical controls, such as biolog-ical controls. These advances are the development of pesticides that degrade rapidly in the environment and the development of improved, more precisely targeted pesticide application techniques. These improvements increase the chances for biological controls to last long enough to be effective.

A focusing of research interest and money occurred during the late 1970s; by 1982, the majority of the USDA research budget for pest control was directed toward biological and nonchemical controls (Batra 1982). Demon-strations of successful research by the USDA, cooperative extension service, and state agricultural experiment stations helped raise farmers' awareness. Training programs and distribution of information by scientists, extension specialists, and pest management specialists led to acceptance of IPM throughout much of the farming community. Evidence of successful IPM systems in local areas has become the best advertisement for this approach.

Presently IPM is being used to varying degrees on the following crops: apples, citrus, cotton, corn, peanuts, sorghum, soybeans, tobacco, trees, vegetables, wheat, and several minor crops. With some crops, such as corn and wheat, IPM practices are very effective; pest losses increase by only an estimated 1-2%. With other crops, such as apples and potatoes, severe, unacceptable losses occur with IPM methods (Pimentel *et al.* 1983). A good historical approach and overview of IPM is available (Poe 1981).

**Practices**

It is not within the scope of this book to go into the specific details of various IPM practices, other than in a general sense. Only one specific exam-ple, involving an IPM program with cotton, is examined, because of its nota-ble success and recentness. Those interested in a more detailed description of specific practices can consult with appropriate extension specialists and the following sources: USDA (1978,1980), Flint and Van den Bosch (1981), and Metcalf and Luckmann (1982).

**Biological**

Pest control in agroecosystems utilizing biological controls is covered by Batra (1982). Based upon her classification, three forms of biological control exist: the classical, augmentative or inundative, and conservative biological controls. Classical biological control utilizes natural enemies of introduced pests. These pathogens, predators, and/or parasites are sought in and intro-duced from the areas from which the original pest was inadvertently import-ed. These enemies must be aggressive, host specific, capable of reproduc-tion, and able to achieve sufficient density to become permanent controls of the pest. Classical biological controls have been successful against weeds and insects. Examples can be found in Caltagirone (1981), Frank (1981), and Batra (1982).

Augmentative or inundative biological control involves the periodic release of imported or native enemies after mass propagation (Batra 1982). These enemies, while they may reproduce under some conditions, are not expected to become permanent controls. The pathogens, predators, and parasites do not necessarily have the host specificity that classical biological controls do. Examples cited by Poincelot (1980) and Batra (1982) include several parasit-ic wasps, an aphid predator *(Chrysopa carnea)*, and several insect patho-gens. *Bacillus thuringiensis* and *B. popillae* are examples of insect patho-gens, in this case against larvae of several moths/butterflies and Japanese beetles, respectively. These particular pathogens are well known commercially.

Conservative biological control utilizes management of all the life forms in an agroecosystem so as to at least conserve, and hopefully to increase, the existing predators and parasites, whether they are native or introduced (Batra 1982). This can be accomplished through use of organic farming tech-niques, polyculture, and stripcropping. The reasons for failure of these natu-ral controls can often be traced to pesticides that suppressed or eliminated beneficial biota.

**Cultural**

Practices involved in organic farming appear highly complementary to conservative biological controls (USDA 1980a). Such organic practices as crop rotation, crop spacing, intercropping, mulching, resistant cultivars, and tillage all have some pest control benefits. As observed in earlier chapters, these practices also offer energy savings and soil improvement. Conven-tional farmers can also manipulate these and traditional practices as part of an IPM program.

Fertilization practices can influence the susceptibility of crops toward dis-eases and pests (Coleman and Ridgway 1983; Hodges and Scofield 1983). In some instances these problems are associated with excessive fertilization and/or nutrient imbalances. Judicious applications in terms of the amount and proportions of nturients, based upon sound soil-testing programs, can often improve IPM results and save money.

Numerous examples exist, but only a few can be cited here. Many can be found elsewhere (Coleman and Ridgway 1983; Hodges and Scofield 1983). Cotton and rice crops are attacked by increasing numbers and types of pests as fertilization (especially nitrogen) increases; a similar situation occurs with cereal crops and leaf pathogens. Mites on apples increase with nitrogen fertilization, but this increase can be countered by raising phosphorus levels. Aphids on tobacco also rise with increased fertilization. Adequate levels of phosphorus decrease fungal infections of wheat. Squash bug feeding on summer squash seems to be associated with inadequate levels of phos-phorus, potassium, and sulfur.

Levels of organic matter in the soil also influence disease and pest suscep-tibility. Nematode populations and root damage appear to decrease as soil organic matter increases (Van der Lann 1956). Several soilborne diseases (see discussion in Chapter 7) are reduced by incorporation of organic matter into soils. Organic matter also stimulates microorganisms known to be disease antagonists. The organic farmer's less than ex-pected problems with insects and disease may well be due in part to his judicious use of organic fertilizers and amendments. Both tend to limit nutrient excesses and imbalances by slow release of nutrients, while maintain-ing organic matter.

**Table 1. Soilborne Palhogens and Diseases Reduced by Organic Maller**

| *Pathogen* | *Crop disease* |
|---|---|
| *Fusarium oxysporum* f. *corianderi* | *Black root of sugar beet* |
| *F. solani* f. *phaseoli* | *Pea root rot* |
| *Phymatotrichum omnivorum* | *Coriander wilt* |
| *Phytophlhora cinnamomi* | *Bean root rot* |
| *Rhizoctonia solani* | *Cotton root rot* |
| Sclerotinia minor | Advocado root rot |
| Streptomyces scabies | Hypocotyl rot of bean |
| Thielaviopsis basicola | Lettuce rot, peanut rot |
| | Potato scab |
| | Bean root rot |
| | Sesame root rot |
| | Tobacco root rot |

*Source*: Adapted from Lumsden et al. (1983)

Most often the decomposition of organic matter reduces diseases, but some instances are known where the opposite is true (Lumsden *et al. 1983).* For example, if plant residues previously infected with a pathogen are incor-porated into the soil, some diseases may be enhanced. Cotton wilt (Ver-*ticillium dahliae)* and bean root rot *(Fusarium soiani* f. *phaseoli)* are examples.

Some cultural practices, such as crop rotations, interplanting, mulching, and time of planting can be complementary to IPM practices (Lumsden *et al.* 1983; Pimentel *et al.* 1983). Rotations of corn with soybean, wheat, and hay are effective for the control of the corn rootworm pest complex. Of these nonhost plants, soybean and wheat are the best choices, as hay crops can introduce several other soil insect pests that attack corn. Rotations of cru-cifers with grains have reduced several vegetable diseases caused by *Rhi-zoctonia soiani* and *Fusarium* spp., especially in the southeastern United States. In the Northwest, snowmold *(Typhuia idahoensis)* can be reduced by rotating wheat with legumes (alfalfa, clover, or peas).

There are some reports that rotations may not be effective for controlling all diseases. Take-all disease of wheat *(Gaeumannomyces graminis)* be-comes less severe with extended monoculture, and more severe when wheat is rotated with legumes or hay. As mentioned previously, hay also increases certain corn pests.

Interplanting can reduce some problems, and increase others, as can changing planting times and using mulches (Pimentel *et al.* 1983). Increas-ing agricultural diversity can reduce insects on certain crops, but the extent to which this occurs is not well documented (Andow 1983). Planting corn early can reduce losses from chinch bug, earworm, fall armyworm, root aphid, rootworm, and southwestern corn borer. However, other pests may become more troublesome and less weed cultivation may be possible. While mulches can control weeds, they may encourage certain pests such as slugs and snails.

Although tillage can provide weed control and the destruction of certain overwintering pests, it tends to increase soil erosion and organic matter oxidation. Minimum and no-till systems reduce the erosion and organic matter problems. The decomposing surface organic matter left in place can control certain diseases, such as stalk rot of sorghum, take-all of wheat, and various soybean rots. However, other disease pathogens are increased: an-thracnose, bacterial blight and wilt of corn, and several soybean diseases (Lumsden *et al. 1983).*

Even irrigation can affect diseases (Hodges and Scofield 1983; Lumsden *et al.* 1983). For example, certain diseases are favored by dry soils, and others by wet soils. Well-irrigated tomato plants can have increased rots from anthracnose and *Rhizoctonia,* but less blossom-end rot. Other pathogens that are less prevalent with heavy irrigation include *Fusarium oxysporum* f. *cubense, Phymatotrichum omnivorum, Sderotinia sderotiorum,* and *Ver-ticillium albo-atrum.*

Careful control of cultural practices can increase the effectiveness of an IPM system. Much of the organic farmer's success without pesticides may be attributed to the conscious and sometimes serendipitous combinations of practices. Our knowledge in this area is incomplete, but the potential for pest control through proper combinations of cultural and biological controls ex-ists. More research is needed to clarify control of specific pests on various

**Table 10.2. Antagonstic Microorganisms Stimulated by Organic Matter**

| *Microorganism* | *Suppressed pathogens or diseases* |
|---|---|
| Actinomycetes<br>Streptomycetes<br>Bacteria | Rhizoctonia solani |
| Pseudomonas eepacia | Fusarium solani,<br>Rhizoctonia solani,<br>Thielaviopsis basicola,<br>Verticillium albo-atrum |
| Fungi<br>Gliocladium roseum | Gaeumannomyces graminis,<br>Rhizoctonia solani |
| Tricoderma hamatum | Rhizoctonia solani |

*Source:* Adapted from Lumsden et al. (1983).crops through the complementary combination of cultural and biological controls.

**IPM with Cotton**

Integrated pest management has been quite successful with cotton and has resulted in decreased usage of pesticides, fertilizers, and irrigation. The cotton IPM system, which is discussed by Adkisson *et al.* (1982), is becoming widely accepted not only because of environmental concerns but because it produces greater net returns than other pest control approaches.

The successful use of IPM in cotton is a significant landmark in the develop-ment of sustainable agriculture. Nearly half of the insecticides utilized in American agriculture are used on cotton. Substantial reduction of national insecticide use would occur if IPM were adopted extensively by cotton growers. In the area of widest adoption, Texas, insecticide use in the late 1970s was only 12% of its highest use in 1964 (OTA 1979).

Five pests are most common in cotton production: the boll weevil *(An-thonomus grandis)*, cotton flea hopper *(Pseudatomoscelis seriatus),* pink bollworm *(Pectinophora gossypiella),* tobacco budworm *(Heliothis vir-escens),* and bollworm *(Heliothis zea).* The first three are primary pests, those that cause the most trouble and reduce profits. The last two are sec-ondary pests; these types only cause problems when their natural enemies disappear.

As insecticide usage in cotton became common during the 1950s, re-sistance among the primary pests increased, causing the need for different insecticides and/ or increased dosages. These changes decimated the natu-ral enemies of the secondary pests, which soon became essentially primary pests. By the late 1960s and early 1970s, problems of resistance peaked. Most registered insecticides for cotton were so ineffective that cotton produc-tion in Texas dropped to one-third of its peak. The stage was set for a new, innovative approach: IPM (Adkisson *et al.* 1982). Ways were sought to mini-mize insecticide usage and maximize control by natural enemies.

A number of elements were combined to produce a very effective (PM system. Conventional long-season cotton cultivars set bolls during peak times for boll weevils, making bolls very susceptive to damage. Therefore, short-season cultivars were developed; these fruited early enough that the boll's carpel was thick enough to resist attack by the time the boll weevils peaked. Further refinements of these cultivars resulted in resistance to seed-ling pathogens, immunity to bacterial blight, and moderate resistance to the cotton flea hopper. One particular cultivar also has moderate resistance to the boll weevil, boll worm, and tobacco budworm.

These cultivars, coupled with certain cultural practices, essentially elimi-nated the need for most insecticides. Shredding and plowing under stalks by mid-September kills most of the diapausing weevils. An insecticide used after harvest further reduces diapausing weevils. Some insectide is used in the spring to control cotton flea hopper and overwintered weevil adults to prevent reproduction. No sprays are needed to control later generations of weevils. This spares the natural enemies of the secondary pests, thus assuring their control. Pink boll worms are controlled by early planting and destruction of crop residues.

The use of a short-season cultivar of cotton also reduces the need for fertilizer and irrigation water. Most cotton farmers in Texas now use some form of this IPM system, which has made it possible for Texas to again produce about 50% of the U.S. cotton crop.

**Economic Benefits**

Environmental and sociological concerns alone would probably not be enough to cause widespread adoption of IPM in the near future. Fortunately, economic benefits also may occur, thus serving as an impetus for adoption and increased research. The road to complete adoption of IPM as a cor-nerstone of sustainable agriculture is not fully paved, but certainly has been begun. The actual extent of present usage is not well documented in the literature.

The extent of economic benefits also is not well-documented. However, the few reported cases strongly suggest dollar savings are quite possible. The best documentation is that by Adkisson *et at.* (1982). Use of lPM systems on cotton, as described in the previous section, resulted in a fourfold increase in net returns per acre compared with conventional pesticide systems. This figure was determined with a well-run demonstration on a private farm where conditions were optimized. Average increases in returns on typical commer-cial farms were not as great, but were certainly notable. Cotton producers on farms in coastal Texas realized increases from $153/ha ($62/acre) with conventional pesticides to $420/ha ($170/acre) with lPM systems. Dryland farmers in Texas had increases of $77/ha ($31/acre). A discussion of the economics of lPM on cotton production in the coastal bend region of Texas is presented by Masud *et at.* (1981).

Another report (Frecon 1983) cites savings of $91 /ha ($37/acre) when an IPM program was used on 2470 ha (1000 acres) of apples and 494 ha (200 acres) of peaches in New Jersey. A second report (Anon. 1983) cites the work of Roger Adams at the University of Connecticut and his lPM system for sweet corn.

This sweet corn project ran for 3 years and involved 10 growers with up to 607.3 ha (1500 acres). This is about 40% of Connecticut's corn land. Actual tests of IPM versus standard pest management were held in sections of each grower's field. Traps utilizing blacklight and pheromones were placed for monitoring the main corn pests: corn earworm (also called cotton bollworm, *Heliothis zea),* European corn borer *(Ostrinia nubilalis),* and the fall ar-myworm *(Spodoptera frugiperda).* Field scouts were also used as monitors. On the IPM plots, pesticide usage declined on the average by 25%;

the high was 50%. This decrease in turn had the indirect benefit of reducing related costs for pesticide usage, such as fuel, equipment maintenance, and labor. Less compaction of soil also was observed, since fewer pesticide field passes were needed. In addition, the corn quality was higher on IPM plots as op-posed to standard practices.

Extensive reviews containing earlier information on the economics of IPM systems are available (McCarl 1981; Osteen *et al.* 1981). Other analyses of the economic potential and impact of IPM strategies can be found in Araji (1981), Musser *et al.* (1981) and Thompson and White (1981).

**Current Research**

Several factors will determine the future development and status of IPM: current and future research; sociological, environmental, regulatory, and economic considerations; changes in technology; improvements in educa-tion and informational dispersal; and farmer acceptance. A number of prob-lems also exist, which mayor may not be resolved. For example, specific IPM systems for crops cannot necessarily be developed on a national scale, since regional differences with pests and other aspects of the agroecosystem exist.

**Research Directions**

Most research tends to focus on specific crops and the pests and other agroecosystems associated with them. The one exception is the gypsy moth *(Porthetria dispar),* the most serious defoliating insect of hardwood trees in the forests, woodlands, and yards of homes in the eastern United States, especially the Northeast.

The USDA and agricultural research stations in the affected states have waged a massive research effort against this gypsy moth. A summary of all the research targeted toward IPM control of this horticultural and forestry pest is covered in excellent detail by Doane and McManus (1981). A proposal for the use of IPM for the cooperative control of the gypsy moth in the most seriously affected states (Connecticut, Maine, Maryland, Massachusetts, New Hampshire, New Jersey, New York, Pennsylvania, Rhode Island, and Ver-mont) can be found in USDA (1981a).

**Much IPM research is targeted toward specific large production crops.**

Recent examples include alfalfa hay (Flint 1981), cotton (Adkisson 1981), rice (Khush and Chandhary 1981), soybeans (Kogan 1981; Carnahan 1982), tobacco (Cheng 1981), and wheat (Nalewaja 1981). Crops of horticultural interest are also the subject of recent IPM research (Wearing 1982). These include apple (Brunner 1981), citrus (Elmer *et aL.* 1981), *GypsophoLa* sp. and chrysanthemums (Price *et at.* 1981), poplar (Filer *et aL.* 1981), potato (LeBrun 1984), deciduous fruits (Fisher and Weinzierl 1981; Mathys 1981), turf (Gibeault *et at.* 1981), and woody ornamentals (Miller and Nielsen 1981). Some lPM research has been targeted toward forest crops (Coulson 1981;

Schmidt and Wilkinson 1981; Ciesla 1982).

A newly emerging research area is that of urban IPM. The control of insects and rodents in inner cities; insects, diseases and weeds in the turf and gar-dens of urban and surburban sites; and insects and diseases on city trees and in parks is beset with difficulties. Diversity is a big problem, whether one thinks of the many pest problems or the different people involved. Further complications exist in terms of economics and personal liberties. Recent reviews on urban IPM research include Carnahan (1982a), Levenson and Frankie (1981), and USDA (1981b).

Another concern of IPM researchers is the influence of environmental conditions and climate on the effectiveness of control by IPM. Coverage of this topic can be found in Duniway (1982), Hatfield and Thomason (1982), and Welch (1982).

Newer vectors of control, to supplement or replace traditional insect pred-ators and parasites and fungal diseases, are receiving attention and may well become the focus of new research. Examples include the use of viruses against insects, allelopathic chemicals against insects and weeds, weeds to serve as sites for beneficial insects and the foiling of pests, and pheromones to suppress insects. The use of insect viruses as control agents has been covered in detail by Payne (1982); Miller *et aL.* (1983) discuss all microbial controls.

Allelopathic chemicals are natural chemical inhibitors released by plants and microorganisms. These chemicals arise from the leaves, litter, and roots in plants and are released into the soil through leaching, decay, and exuda-tion. They are of particular interest because of their herbicidal properties. Interestingly, some of the crops used in rotation schemes on farms show strong allelopathic traits. These crops include oats, soybeans, sorghum, wheat, rye, barley, and sunflower. These particular crops appear to suppress weed growth for succeeding crops not showing strong allelopathic charac-teristics. Weed control through the use of annual rotations or companion plantings of allelopathic crops may become a future agricultural practice. An excellent discussion of allelopathic chemicals is presented in Putnam (1983).

Normally, the IPM approach toward weeds has been to reduce their growth or eliminate them. Recently, attention has focused on the use of certain weeds as hosts for beneficial species, as repellants against harmful insects and nematodes, and as decoys or traps for undesirable insects and patho-gens. An excellent summary of these approaches can be found in William (1981).

Controlled release of pheromones, natural sexual attracts of insects, to suppress insects has received widespread attention. Many home owners have utilized pheromone traps against gypsy moths and Japanese beetles. Some pheromones have been used successfully with major crops, such as cotton. Detailed, comprehensive coverage of this topic is available (Ky-donieus and Beroza 1982).

## Future Directions

A number of issues are in need of further resolution. For example, the methodology to predict pest arrivals, population levels, and damage conse-quences needs refinement. One approach possible is systems modeling, which is based upon mathematical modeling or simulations of crop growth and pest dynamics with the assistance of computers. While theoretical, prop-er development of models can produce an operational approach fully capa-ble of providing the predictions required for success with IPM. Crop scouting, a practical tool, is used for assessment of crops and pests. It provides an ongoing informational basis for IPM use, as opposed to the previous predic-tive informational base. Both are needed for good results with IPM systems. Someday, the ongoing information collected by scouts may be transmitted via computer links directly to the farmer. One such program like this almost exists in Kentucky. Almost, because presently the information exists on the computer in the county extension office and the farmer calls in to get it (Brown 1982). The obvious future is computer access directly from the farmer's home computer.

Modeling presently plays some role in IPM. For example, some existing models cover the dynamics of pest populations for specific crops and pests. Covered crops include alfalfa, cotton, forests, peanuts, and soybeans (Rue-sink 1976). Some models of plant pathogens (Tummala *et al.* 1976) are available: apple scab *(Venturia inaequaLis)*, potato blight *(Phytophthora infestans)*, and wheat stem rust *(Puccinia graminis tritici)*. Hartstack and Witz (1981) discuss modeling of alfalfa weevil *(Hypera postica)*, boll weevil *(Anthonomus grandis)*, earworms *(HeLiothis)*, Lygus bugs *(Lygus hesperis)*, and mites. To date, models have concerned major pests or crops, where economic losses can be substantial; expansion of models to other crops and pests is needed.

Modeling is also the working tool of system science, which is and will be increasingly applied to agriculture, especially IPM. Agricultural scientists ac-cept the systems concept, which basically seeks to resolve complex agroeco-systems into various components and subsystems. Modeling is used to provide a working facsimile of each component. Each component interacts with the other to make up the system. The model describes the component and its relationship to others. These models can be manipulated to predict the behavior of the system, given different inputs. A good discussion of the basic mathematics and computer input of agricultural modeling is available (DeMi-chele 1975).

A number of factors are involved as components in the system. These include crop development, pests and pest enemies, cultural factors, and several environmental factors (e.g., humidity, nutrients, rain, soil, tem-perature, and water): These components may either be controlled or uncon-trolled factors. Controlled factors can include, for example, cultural practices and cultivar qualities, whereas uncontrolled factors include pests, tem-perature, and nutrients.

Since uncontrolled factors change with time, they require frequent updat-ing to truly reflect the current status of the system. If done, the model system becomes a working tool for ongoing decision making and adaptibility to changing conditions. Such an apprJach is called on-line pest management (Tummala 1976).

Used in this manner, system science application to pest management or other agricultural practices offers several uses. It can provide resolution and insights into complex agroecosystems, identify specific areas as factually weak and serve to show research priorities, predict behavior and serve to assess needs of pest management and other cultural practices, and serve to provide data for interdisciplinary research needs.

Much more research is needed on system science for IPM. Presently, large-scale models exist for forestry and cotton (Ruesink 1976), two crops that now face increasingly serious pest threats. Conventional pest management ap-pears less desirable for these crops because of rising costs, increased pesticide resistance, the need for increasingly stronger pesticides at higher dosages, and rising environmental concerns.

The other models discussed earlier are not for the total system, but are for some of the system components. More modeling is needed to complete those systems, and the system science approach needs to be extended to presently uncovered crops. Present crops that depend heavily on modeling in their IPM programs include apple, celery, citrus, corn, cotton, and soybeans (Poe 1981).

Modeling systems also need to accommodate regional and even local differences in IPM programs. This undoubtedly will involve extension agents and some computer technology at the regional and even farm level. Accep-tance of computer technology at the farm level will undoubtedly happen at corporate farms and to a degree at single family farms (see Chapter 10). Consultants, who already supply much IPM advice, may also supply comput-er information at local levels.

In conclusion IPM will remain and expand. More research is needed at the university and state experiment station level, cooperation is needed on the regional level for planning and coordination, and education must supply the professional consultants, knowledgeable extension agents, and aware farm-ers. The USDA commitment of extensive funds to IPM research, regional programs, and education is encouraging (Kuhr 1981; Batra 1982).

### Genetic Resources

Germplasm is a very critical agricultural resource which receives little at-tention. Crop productivity increases through plant breeding have been and will be dependent upon germplasm resources. The lack of attention results from germ plasm's low visibility as a resource. Germplasm is not observable like soil and water, nor is the erosion of genetic resources as obvious a problem as soil erosion or water loss.

We know that successful plant breeding is highly dependent upon the worldwide availability of genetic diversity. Plant breeders use germplasm for a number of plant improvements. One important use is the introduction of cultivars showing first time resistance towards an existing disease or insect. Equally important is the reintroduction of resistance in plants where modified strains of diseases and insects have managed to circumvent existing re-sistance. Other new cultivars are bred to satisfy changes in cultivation or market demands. Improved production of natural products from crops, such as medicinals or substitute fuels, is also dependent upon the diversity of genetic resources.

Genetic diversity is being lost at an alarming rate outside the United States (OTA 1982a). Losses are especially heavy in tropical forests as a conse-quence of increased urbanization, industrial development, and agricultural production. Estimates place the current total loss of tropical forest cover at 40% and the annual clearance loss continues at 1-2%. Unless abated, trop-ical forests will be 75% lost by the year 2000. While assessment is difficult, the remaining tropical forests are extrapolated to contain nearly one half of the earth's plant species. The majority of these species are unknown in terms of identity and value as a genetic resource. Estimates place the ongoing yearly plant extinctions at several hundred in addition to thousands of indige-nous wild crop types lost already. Once lost, germplasm is irreplaceable.

Solutions to the problem are not easy, since costs are high and the im-pacted areas are beyond the United States (OTA 1982a). Clearly germplasm repositories in the United States must be enlarged and broadened. Collection efforts need to be increased. Equally essential is the preservation of natural genetic resources until scientists are able to complete the identification, collection and evaluation of the sources of genetic diversity. Options concerning germplasm have been presented to Congress by the Office of Technology Assessment.

A related threat concerns genetic vulnerability. Our major crops now have a narrow genetic base as a consequence of inbreeding resulting from the needs of farmers, food processors, and consumers. This selection for and national dominance by a few best cultivars for each crop is potentially dan-gerous. Should a disease mutation arise and infect a major crop, severe damage could result for that crop. The devastating corn blight in 1970 raised the specter of food crop susceptibility to severe disease outbreaks.

Increased genetic diversity in our major crops is desirable and possible, but social, economic and free market pressures complicate the issue. This area clearly needs more attention. At the least (OTA 1982a), an early warning system for the recognition of potential vulnerability of crops should be developed.

## REFERENCES

Adkisson,, P. L., Niles, G. A., Walker, J. K., Bird, L. S. and scon, H. B. 1982. Controlling cotion's insect pests: A new system. Science 216, 19-22.

Alberts, E. E., Schuman, G. E. and Burwell, R. E. 1978. Seasonal runoff losses of nitrogen and phosphorus from Missouri *Valley* loess watersheds. J. Environ. Qual. 7 (2), 203-208.

Allen, G. E. 1980. Integrated pest management. Bioscience 30, 655-701.

Andow, D. 1983. Effect of agricultural diversity on insect populations. *In* Environmentally Sound Agriculture. W. Lockeretz (Editor). Praeger Publishers, New York.

**Anon.** 1983, Connecticut pest management with IPM. Amer. *Veg.* Grower 31 (4) 46. ARAJI, A. A. 1981. The economic impact of investment in integrated pest management. Res. Bull. *115.* Idaho Agric. Exp. Stn.

Batra, S. W. T. 1982. Biological control in agroecosystems. Science 215, 134-139. BRADFORD, R. R. 1974. Nitrogen and Phosphorus Losses from Agronomy Plots in North Alabama. EPA-660/2-74-033. U.S. Environ. Protection Agency, Washington DC.

Brown, G. C. 1982. Microprocessor-based information management system for an inte-grated pest management system. Bull. Entomol. Soc. Amer. 28, 135-137.

Brunner, J. 1981. New tactics for apple IPM. pp. 145-150. *In* Proc. Washington State Horticultural Assoc. 1981.

Caltagirone, L. E. 1981. Landmark examples in classical biological control. Annu. Rev. Entomol. 26, 213-232.

Carnahan, W. B. 1982a. Urban IPM: Blooming across the country. Extension Rev. 53 (2), 12-14.

Cheng, H. H. 1981. Progress in integrated pest management of tobacco insects in Canada. The Lighter 51 (3) 17-21.

Ciesla, W. M. 1982. IPM: New approaches to old problems. Amer. For. 88 (2) 40-44, 51-52.

Coleman, E. W. and Ridgway, R. L. 1983. Role of stress tolerance in integrated pest management. *In* Sustainable Food Systems. D. Knorr (Editor). *AVI* Publishing Co., West-port, CT.

Coulson, R. N. 1981. Evolution of concepts of integrated pest management in forests. J.

Ga. Entomol. Soc. 16 (Suppl. 1),301-316.

Demichele, D. W. 1975. An evaluation of modelling, systems analysis, and operations reserach in defining agricultural research needs and priorities in pest management. Iowa State J. Res. 49, 597-621.

Doane, C. C. and McManus, M. L. 1981. The Gypsy Moth: Research Toward Integrated Pest Management. Tech. Bull. U.S. Dept. Agric. Forest Service, Washington, DC.

Duniway, J. M. 1982. Role of biometerology in integrated pest management: **Soil, plant, water** relations and disease. *In* Biometeorology in Integrated Pest Management. Academic Press, New York.

Eelmer, H. S., Brawner, O. L. and Ewart, W. H. 1981. The citricola scale in San Joaquin IPM programs. Citrograph 66 (6) 137-138, 151.

EPA. 1982. Water Quality Assessment: A Screening Procedure for Toxic and Conventional Pollutants. EPA 600/6-82-004a-c. Center for Environ. Res. Information, U.S. Environ. Pro-tection Agency, Cincinnati, Ohio.

Filer, T. H., JR., Solomon, J. D., Cooper, D. T. and Hubbes, M. 1981. Integrated pest management of poplar species. *In* Proc. IX Intern. Congo Plant Protection, 1979. Burgess Pub., Minneapolis.

Fisher, G. C. and Weinzierl, R. 1981. Integrated Pest Management and Its Potential Application in Small Fruits Production. Annu. Rept. Oregon Horticultural Soc., Vol. 72.

Flint, M. L. 1981. Integrated Pest Management for Alfalfa Hay. Statewide IPM Project. IPM Manual Group. University of California, Berkeley.

Flight, M. L. and Van Den Bosch, R. 1981. Introduction to Integrated Pest Management. Plenum Press, New York.

Fox, J. L. 1983. Soil microbes pose problems for pesticides. Science 22 1, 1029-1031. FRANK, J. R. (Editor) 1981. Integrated pest management: Present and future. Proc. of symposium, sponsored by Weed Control and Pest Management Working Group of the Ameri-can Society for Horticultural Science. HortScience 16, 500-516.

Frecon, J. 1983. IPM saves money. Amer. Fruit Grower *103* (1) 19.

Friend, G. 1983. The potential of an alternative agriculture. *In* Sustainable Food Systems.

D. Knorr (Editor). AVl Publishing Co., Westport, CT.

Gibeault, V. A., Bowen, W. R., OHR, H. D., Thomason, I. J. and Cress, F. 1981.

Integrated pest management for turf. Calif. Turfgrass Culture 31 (2) 13-15.

Hanson, D. J. 1984. Agricultural uses of ethylene dibromide halted. Chem. & Eng. News (March 5), 13-16.

Hanthorn, M. and Duffy, M. 1983. Returns to Corn Pest Management Practices. Agric.

Econ. Rept. No. 501. U.S. Dept. Agric., Washington, DC.

Hartstack, A. W. and WITZ, J. W. 1981. Insect modeling. Agric. Eng. 62 (9) 19-20. Hatfield, J. L. and Thomason, I. J. (Editors). 1982. Biometerology in Integrated Pest Management. Academic Press, New York.

Hodges, R. D. and SCOFIELD, A. M. 1983. Effect of agricultural practices on the health of plants and animals produced: A review. *In* Environmentally Sound Agriculture. W. Lock-eretz (Editor). Praeger Publishers, New York.

Khush, G. S. and CHANDHARY, R. C. 1981. Role of Resistant Varieties in Integrated Pest Management of Rice. Extension Bull. No. 162. Food and Fertilizer Technology Center, Taipei City, Taiwan.

Klein, D. T. and KLEIN, R. M. 1985. The growing case against acid rain. Garden 9 (2) 22-27.

Kogan, M. 1981. Dynamics of insect adaptations to soybean: Impact of integrated pest management. Environ. Entomology 10 (3) 363-371.

Kress, L. W. and Miller, J. E. 1983. Impact of ozone on soybean yield. J. Environ. Qual. 72 (2) 276-281.

Kuhr, R. J. 1981. Regional planning and coordination of integrated pest management pro-grams. HortScience 16, 514-515.

Kydonieus, A. F. and BEROZA, M. 1982. Insect Suppression with Controlled Release Pher-omone Systems. Vols. 1 and 2. CRC Press, Boca Raton, FL.

Lebrun, R. A. 1984. Will *B.b.* beat the beetle? Amer. Veg. Grower 32 (1) 8.

Levenson, H. and Frankie, G. W. 1981. Pest control in the urban environment. Prog. in Resource Management and Environ. Planning 3, 251-272.

LUMSDEN, R. D., LEWIS, J. A. and Papavizas, G. C. 1983. Effect of organic amendments on soilborne plant diseases and pathogen antagonists. *In* Environmentally Sound Agri-culture. W. Lockeretz (Editor). Praeger Publishers, New York.

Masud, S. M., Lacewell, R. D., Taylor, C. R., BENEDICT, J. H. and L1PPKE, L. A. 1981. Economic impact of integrated pest management strategies for cotton production in the coastal bend region of Texas. South. J. Agric. Econ. ]3 (2) 47-52.

Mathys, G. 1981. Implementation of integrated pest management on deciduous fruits in Europe. *In* Proc. IX Intern. Congo Plant Protection, 1979. Burgess Pub., Minneapolis.

McCarl, B. A. 1981. Economics of integrated pest management: An interpretive review of the literature. Oregon Agric. Exp. Stn. Special Rept.

Metcalf, R. L. and LUCKMANN, W. H. 1982. Introduction to Insect Pest Manatement. 2nd ed. John Wiley & Sons, New York.

Miller, J. E. 1983. Studies reveal effects of air pollution on crops. Logos] (3) 20-23. MILLER, L. K., LINGG, A. J. and BULLA, L. A. 1983. Bacterial, viral and fungal insecticides. Science 2] 9, 715-721.

Miller, R. L. and NIELSEN, D. G. 1981. Insect and Mite Control on Woody Ornamentals. Ext. Bull. 504. Ohio State Univ.

Musser, W. N., TEW, B. V. and EPPERSON, J. E. 1981. An economic examination of an integrated pest management production system with a contrast between E-V and stochas-tic dominance analysis. South. J. Agric. Econ. ]3 (1) 119-124.

Nalewaja, J. D. 1981. Integrated pest management for weed control in wheat. *In* CRC Handbook of Pest Management in Agriculture, Vol. 3. CRC Press, Boca Raton, FL.

Osteen, C. D., BRADLEY, E. B. and Moffitt, L. J. 1981. The Economics of Agricultural Pest Control: An Annotated Bibliography 1960-80. U.S. Dept. Agric. Economics and Statis-tics Service.

OT A. 1979. Pest Management Strategies: Present and Future Pest Management Strategies in the Control of Sorghum and Cotton Pests in Texas. Office of Technology Assessment, Washington, DC.

OT A. 1982. Impacts of Technology on U.S. Cropland and Rangeland Productivity. Office of Technology Assessment, Washington, DC.

OTA. 1982a. Genetic Technology: A New Frontier. Office of Technology Assessrnent. West-view Press, Boulder, CO.

Parr,J. F., Marsh, P. B. and KLA,J. M. 1983. Land Treatment of Hazardous Wastes. Noyes Publications, Park Ridge, NJ.

Payne, C. C. 1982. Insect viruses as control agents. Parasitology 84 (4) 35-77. PIMENTEL, D. and EDWARDS, C. A. 1982. Pesticides and ecosystems. Bioscience 32, 595-600.

Pimentel, D., Glennister, c., FAST, S. and Gallahan, D. 1983. An environmental risk assessment of biological and cultural controls for organic agriculture. *In* Environmentally Sound Agriculture. W. Lockeretz (Editor). Praeger Publishers, New York.

Poe, S. L. 1981. An overview of integrated pest management. *In* HortScience ] 6, 501-506.

Poincelot, R. P. 1980. Plant protection. *In* Horticulture: Principles and Practical Applications. Prentice-Hall, Englewood Cliffs, NJ.

Price, J. F., Englehard, A. W., Overman, Q. W., Yingst, A. J. and IVERSON, M. K. 1981. IPM demonstrations in commercial gypsophola and chrysanthemums. *In* Proc. of the 1980 Meeting, Florida State Horticultural Society.

Putnam, A. R. 1983. Allelopathic chemicals: Nature's herbicides in action. Chem. Eng. News (April 4), 34-45.

Ruesink, W. G. 1976. Status of the systems approach to pest management. Annu. Rev.

Entomo!. 2], 27-44.

Schmidt, R. A. and Wilkinson, R. C. 1981. Prospects for integrated pest management in slash pine ecosystems in Florida. *In* Proc. IX Intern. Congo Plant Protection, 1979. Burgess Pub!., Minneapolis.

Thompson, P. and WHITE, G. B. 1981. An Economic Evaluation of the Potential for Tree Fruit Integrated Pest Management in the Northeast. Dept. of Agricultural Economics, Cor-nell Univ. Agric. Exp. Stn., Ithaca, NY.

Tummla, R. L., Haynes, D. L. and CROFT, B. A. 1976. Modeling for Pest Management:

Concepts, Techniques and Application. Michigan State Univ., E. Lansing.

USDA. 1978. Biological Agents for Pest Control: Status and Prospects. U.S. Dept. Agric., Washington, DC.

USDA. 1980. Basic Principles of Insect Population, Suppression and Management. Agric.

Handbook 512. U.S. Dept. Agric., Washington, DC.

USDA. 1980a. Report and Recommendations on Organic Farming. U.S. Dept. Agric., Washington, DC.

USDA. 1981. America's Soil and Water: Condition and Trends. U.S. Dept. Agric., Washington, DC.

USDA. 1981 a. Cooperative Gypsy Moth Suppression and Regulatory Program Activities. U.S. Dept. Agric., Forest Service, Washington, DC.

USDA. 1981 b. Urban Integrated Pest Management: A Report. U.S. Dept. Agriculture, Science and Education Administration. Extension Committee on Organization and Policy. Univ. of Georgia, Athens.

Van Der Lann, P. A. 1956. The influence of organic manuring on the development of the potato root eelworm, *Heterodera rostochiensis.* Nematology ], 113-125.

Weairing, C. H. 1982. Integrated pest management: Progress and prospects with special reference to horticulture. N. Z. J. Exp. Agric. ]0 (1) 87-94.

Welch, S. M. 1982. Risk-based design of meteorological networks for integrated pest management. *In* Biometeorology in Integrated Pest Management. Academic Press, New York.

William, R. D. 1981. Complementary interactions betwen weeds, weed control practices, and pests in horticultural systems. HortScience ] 6, 508-513.

# 11

# Agriculture Biotechnology

Over the last two hundred plus years, American agriculture has responded to the various demands and problems thrust upon it. The success of that response owed much to research, and the development and implementation of new technologies. Among the most noteworthy technologies have been chemical aids, such as fertilizers, pesticides, and growth regulators; labor-saving machines, such as tillage equipment, planters, applicators for pesticides and fertilizers, and harvest equipment; and new cultivars designed for; naximal yield, pest resistance, and compatibility with existing technologies.

Once again American agriculture is faced with serious demands and prob-lems, some of which result from previously adopted technology. Much of the technology depended on inexpensive energy, but now energy consumes a painful part of the agricultural budget. Chemical aids, because of their energy costs and environmental impact, are no longer the great panacea. In an effort to maintain profits in the face of escalating costs, some farmers neglect the maintenance of their basic resources, leading to diminished soil and water quality.

As these concerns mounted, the new direction for American agriculture became one of sustainability. With awareness came responses, such as those discussed in previous chapters. Technologies associated with prob-lems were not discarded but modified to meet new demands. Chemical pesticides blended with nonchemical pest controls to become a newer tech-nology, integrated pest management. Chemical fertilizers were joined to soil-conditioning methods favored by organic farmers. Energy use was made more efficient and better management of soil and water was started.

The objective of a sustainable agriculture has not yet been achieved; much remains to be accomplished. Most of the future progress in achieving a sustainable agriculture will rely on adoption of newly emerging technology and the creation of even newer forms. Key parts of this new technology will be computer technology, solar technology, and biotechnology. Surveys of agriculture's possible future are available from the Futures Group (1982) and the Battelle Memorial Institute (1983).

Before moving on to a discussion of new technologies for sustaining agriculture, it should be pointed out that substantial research remains yet on existing technologies. These areas were highlighted throughout the preced-ing chapters. This near-term and future long-term research will require sub-stantial federal funding if a sustainable agriculture is to become a reality. In view of the consequences if we fail, such funding must be given a high priority.

WIth this need in mind, it is interesting to note that the President's Private Sector Survey on Cost Control has reached some interesting conclusions in its 1983 draft report on the Department of Agriculture. The possible savings have been grouped into various categories, based upon the degree to which they are realistically substantiated, defensible, and supportable on their man-agement merits. The most realistic recommendation, one which is fully sub-stantiated and defensible, is that some $10 billion cost savings are possible for the Department of Agriculture over 3 years. While social, economic, and/political factors will influence the implementation of these savings, serious! thought should be given to these recommendations as a possible way to fund the nation's future in sustainable agriculture.

## Solar Technology

Solar energy, technology and economics being favorable, can presumably be substituted for stationary fuels used on the farm. Such fuels are electricity, natural gas, and LP gas; these energy sources are used on the farm and residence for heating, cooling, drying, lighting, and running various pieces of equipment and appliances. These fuels account for about 22% of farm energy needs. An average U.S. farm requires about 934 million Btu of energy each year; 84% goes for business operations and 16% for the residential parts of the farm (Table 11.1).

**Table 11.1: Farm use of Stationary Fuels**

| | *Usage by fuel (Btu × $10^9$)* | | |
|---|---|---|---|
| Activity | Electricity | LP gas | |
| Crop drying | | 57904 | 700 |
| Home heating | 2350 | 29740 | 36680 |
| Livestock maintenance | 21165 | 5001 | — |
| Poultery breeding | | 19835 | 700 |
| Water heating | 2220 | 2100 | 2480 |
| Total (Btu x $10^{12}$) | 1136.4 | 119.6 | — |

Fuels likely to be replaced by solar energy are electricity and LP gas, mostly because of costs. With changing inflation and decreasing supplies, it is not easy to predict long-term fuel cost differentials. The present national cost ratio of electricity: LP gas: natural gas per 100,000 Btu is about 3.2: 1.4: 1. Natural gas presently costs less than solar energy, so it is not likely to be replaced by solar energy unless deregulation and supply problems drive the price of gas upwards. A general discussion on solar energy is presented by Spillman (1981).

## Crop Drying

The energetics of crop drying were discussed in Chapter 3. About 90% of the crop-drying energy is used on corn. Other dried crops include rice, hops,

hay, peanuts, tobacco, and some fruits and vegetables. The present outlook *for* solar technology applications to crop drying is quite good. With good. Designs and multiple-use efficiency, the cost payback period is 3.5-5 years, making solar technology economically feasible (Heid and Trotter 1982).

More research is needed on portable, multiple-use designs. Most of the search in the 1970s produced stationary solar systems designed for drying a specific crop. Systems were usually air-type, flat-plate collectors. Present research is moving toward portable, multiple-use collectors, since ducting to sites and buildings can result in heavy heat loss at 15.2-30.5 m. (50-100 ft). However, multiple uses entail variable heat requirements and air flow. Such variability makes it difficult to obtain excellent efficiency under all con-ditions, thus making some compromise necessary.

Research on solar driers for grain, mostly corn, has produced a number of conclusions. Air-type collectors were shown to be more efficient than liquid types with heat exchangers. In terms of economics and simplicity of on-farm construction, they were deemed better than concentrating high-temperature collectors. Drying of high-moisture grain was not profitable, since stirrers and a back-up system were required. The latter was necessary for periods of low solar radiation and high humidity (Thompson and Pierce 1977; Heid and Aldis 1981). Examples of collector designs in the 1970s are discussed by Heid and Trotter (1982).

**More recent examples of solar grain-drying designs are quite promising.**

One collector, designed both for crop drying and heating of livestock quar-ters, showed economic competitiveness with conventional energy (Heid and Trotter 1982). Another has multiple uses, portability, flexible airflow, and tiltability. Payback time was estimated at 5.8 years (Heid 1981). Plans are available from the University of Illinois (1979) for a multiple-use, portable collector which could be built for around $1300 to $1400 at today's costs. Interest and acceptance were great enough for the USDA to start a solar grain-drying demonstration project in 1980 (Heid and Trotter 1982).

Other solar designs have been researched for drying of hay, hops, fruits and vegetables, as well as the curing of peanuts and tobacco. Again the short period of drying activity makes solar drying uneconomical for commercial applications unless multiple uses are considered.

For example, hay or hop drying is done during the summer. This would leave the solar drier free for use in grain drying and space heating applica-tions. Another possibility is for the sequential use of a drier on tobacco, corn, and then peanuts. Another is for solar energy to be used in a greenhouse for heating of crops and for curing of peanuts and tobacco (McElroy and Krause 1982).

Recent solar-drying research on hay includes that of Bledsoe and Henry (1981) and Morrison and Shove (1981). Drying of hops with solar collectors has been examined by Kranzler (1981), and of vegetables and fruits by Coleman (1981). The curing of peanuts with solar driers was studied by Schlag and Sheppard (1979) and Butler and Troeger (1981), and of tobacco by Cundiff (1981), Henson (1981), and Huang and Toksoy (1981). Guceri (1983) has recently reviewed heated air solar collector use with various crops.

An interesting departure from the heated air-type collectors for drying grain, using a pond as a solar collector, shows promise. The solar pond concept is presently an active research project at the Argonne National Labo-ratory (1983). The key to the solar pond is salt water. The salt concentration increases with depth, forming gradients that minimize convection currents in the pond. The heavier layers are heated by the sun's rays. Since the layers cannot rise, the heat is trapped. The pond reaches temperatures over 77°C (170°F) in the summer, and remains above 38°C (100°F) in the winter. The temperature peak is reached in September, which is ideal for the grain-drying season. More work is needed on the heat extraction equipment for such a system. The solar pond concept has also been examined at the University of New Mexico, which has one of the oldest projects, at Ohio State University, and in Israel (SERI 1981a). Solar ponds can also be utilized for space and water heating, a topic which we will now examine.

## Space and Water Heating

The heating of farm space and water runs a close second to crop drying in terms of solar potential. Some possible uses in this category include the heating of various farm buildings, such as animal quarters and the residence, and the heating of water for these same applications. Solar units for these purposes can be for a single use and still be economical, assuming efficient designs.

The most promising solar uses, in terms of design and ongoing research, are for heating of swine and poultry houses, and providing hot water for dairy operations. This is especially true when solar collectors are used both to heat and cool buildings. Under these conditions a payback period of 5 years is likely and was achieved on some 90 solar demonstration livestock farm projects. A number of these designs for swine, poultry, and dairy opera-tions are discussed by Heid and Trotter (1982).

One system can be used year-round in swine houses for farrowing and nursery purposes (Houghton 1980). The solar collector provides both heat and hot water; the payback time appears to be near 5 years. This system is now commercially produced.

Space heating for the farm residence (or many homes for that matter) also appears to be economically feasible, especially in view of energy tax credits. Some systems are already available on the market, and a number of do-it-yourself designs exist. A similar situation also exists for solar hot water sys-tems. Heid and Trotter (1982) discuss both systems. A number of residential solar-heating systems are covered in federal publications (HUD/DOE 1979; SERI 1981a).

A number of research projects have examined the use of solar energy in several other areas. These have potential applications in agriculture, but do not presently have as great an economic feasibility or promise as the uses already discussed. This may change if energy costs continue to rise. These areas include coolers, di§tillers, greenhouses, and irrigation. A discussion of these solar energy applications can be found in Heid and Trotter (1982).

A marketing analysis on present and projected solar energy technology was performed by Heid and Trotter (1982), who indicated a promising future for agricultural use, provided that more research and development were conducted to refine the technology. Improved communications and better delivery of the technology to farm users is also a prerequisite. One problem is that agriculture needs low-cost technologies, which take second place to high-cost technologies in terms of marketing and development by the solar industry. Declining federal funding for agricultural applications further com-plicates the problem. Even so the Department of Energy estimates that by 2000 half of the energy used in agriculture could be replaced by solar energy.

One exception to the above appears to be the marketing and development of photovoltaic systems. As pointed out in Chapter 3, photovoltaic systems presently have some limited use in irrigation systems in the Southwest. Since 8% of all the energy use on the farm is electricity, many other uses for electricity from photovoltaic systems are possible. Besides irrigation, uses in the near future include applications in poultry farming and electric fences. Projections are that solar-produced electricity may become competitive with conventional electricity by the late 1980s (SERI 1981a).

**Windpower**

Many of the basic practices being adopted to form today's sustainable agriculture were commonplacc on farms in the 1800s. One of these was the use of windmills to pump water and generate electricity. The wind is convert-ed to electricity at an efficiency comparable to fossil fuel power plants. A discussion of windpower (Fig. 10.3) is provided by Klueter (1981).

The key to the economic feasibility of windpower in today's agriculture is similar to that with solar energy-multiple uses. Suggested uses of the gen-erated electricity include heating of buildings, crop drying, irrigation, and cooling operations. The economics of wind energy is improved if the system can store electricity for use during windless times and if surplus energy can be sold to utilities through an electrical grid system.

Areas of sufficient wind to provide reasonable energy exist in about one-half of the United States. Very few of the favorable areas exist in the eastern United States. Even in favorable areas, the potential contribution of windmills is thought of as supplementary to solar energy systems. Additional informa-tion on wind energy, appropriately enough in view of the foregoing projec-tion, is available from the Solar Energy Research Institute (SERI 1980a), and also from NASA (1982).

The use of waterwheels to do work and generate small amounts of elec-tricity is another old practice, once very common in the 1800s. Today's successor, various forms of water turbines, may also increase production of energy on the farm.

Areas in the eastern United States, unable to generate wind energy, are fortunate in that they have many potential hydroelectric sites. Other favorable areas include the Northwest and northcentral states. More than likely, some farms in these areas with favorable hydroelectric sites will derive some of their future energy needs from hy-droelectric power.

One particular advantage is that low-head hydrogeneration is constant and con-trollable, a distinct advantage over wind energy. Sales of surplus energy through an electric grid are probably more realistic because of this advan-tage. The initial cost and possible legal problems are presently a drawback. A good discussion on hydroelec-tric power is presented by Matson (1981). Information on construction costs and production expenses is available (DOE 1983).

Geothermal energy can be used to heat buildings and provide energy for other purposes. Some present uses are the heating of homes and buildings in Idaho and Texas, provision of electric power in California, the drying of onions in Nevada, and the raising of fish in Idaho. The major hot ground-water resources for geothermal energy tapping are in the western United States, but some sites do exist in eastern and western coastal areas. Abel and Walker (1981) discuss geothermal energy.

Potential uses of geothermal energy on farms include the heating of greenhouses, homes, animal facilities, and ponds used for aquaculture. A number of agricultural applications are currently in use. For example, a ranch in South Dakota derives heat for buildings, grain drying, and water warming for livestock. Fish farming in Califor-nia, Colorado, Idaho, Oregon, Utah, and Wyoming take advantage of geothermal resources.

Some food processing in Nevada utilizes geothermal energy. Undoubtedly some farms will provide some of their future energy needs through geothermal systems.

The sustainable agricultural futurist gets especially excited when certain aspects of biotechnology are examined: genetic engineering and biomass. Biotechnology may be the salvation of agriculture someday. The keyword is "someday." Our present knowledge about the molecular biology, bio-chemistry, genetics, and physiology of plants is not so complete as we would like to think; much more data is needed to assure the successful develop-

ment and application of biotechnology to agriculture. Commerical applica-tions of biotechnology will probably not be realized until the 1990s. Until then we must depend upon current technologies to sustain agriculture, such as energy conservation and IPM. Still the consensus is favorable for future help from biotechnology.

**Genetic Engineering**

The manipulation of plants (and animals) to suit agricultural needs has its basis in genetics. The accomplishments of applied genetics during the past several decades have depended on classical genetic methods, that is, con-ventional breeding and se-lection for desired characteristics. The successes of this approach are documented in a recent article by Borlaug (1983).

The newer approach to improving plants has its basis in molecular genet-ics, that is, direct manipulation or engineering on the molecular and cellular levels using techniques such as recombinant DNA, cell fusion, tissue culture, and *in vitro* fertili-zation. The new genetic engineering should be thought a.f as a complement or supple-ment to conventional plant breeding, not as a replacement. A danger exists that the excitement and promise of genetic '-engineering may lure researchers and funding completely away from con-ventional plant breeding, which still has much to offer sustainable agriculture (Borlaug 1983). An excellent overview of genetic engineering is available (OTA 1982).

The recombinant DNA approach involves the splicing of different pieces of DNA together and then their introduction into a host. One piece of DNA contains the gene or genes that can express a desirable genetic trait, such as disease resistance or nitrogen fixation. These genes can be selected from any species; the constraints of natural breeding barriers are bypassed. For example, genes from bacteria, other plant species, or even an animal could be introduced into the recipient host, whether it be a plant, animal, or micro-organism. The second piece of DNA serves as a carrier or vector. The desirable DNA is attached to the vector, which then transfers the useful, but foreign, DNA into the host's genetic material. Another characteristic of the vector is that its genes are read by the host and then expressed. If done correctly, the genes added to the vector will also be read and expressed. The DNA is introduced into a cell, which must then be regenerated into an organism.

Using this recombinant DNA method, scientists have produced two vac-cines that are presently undergoing tests and may appear on the market in the late 1980s. One is a vaccine to prevent a specific strain of hoof and mouth disease in cattle; the other prevents colibacillosis, an intestinal swine infection (McElroy and Krause 1982). Another successful example of the recombinant DNA method is the introduction of antibiotic resistance from a bacterium into a petunia plant (NRC 1984).

Other applications of recombinant DNA, utilizing microorganisms as host cells, also offer promising results for agriculture. These include production of other vaccines, antibiotics, and hormonal growth stimulators. The latter en-hance milk and meat production and will be used with cattle, poultry, and swine. Preliminary tests for milk production are good. Microorganisms al-tered by recombinant DNA may also produce essential amino acids, vi-tamins, and otl'ler additives for use as supplements for animal feeds, or even convert biomass into food for humans and animals. Further off is also the possibility that this approach could be used to produce pesticides from host cells (Miller *et aL.* 1983). Many other agricultural uses are also envisioned further in the future (McElroy and Krause 1982). One interesting future application of genetic engineering concerns the issue of diminished germ-plasm resources. Since molecular engineering has the potential to introduce specific traits from plants outside of compatible breeding populations, it could partially offset losses in genetic diversity. Barton and Brill (1983) pro-vide an interesting discussion on future prospects and the numerous prob-lems impeding progress.

Two problems impose substantial limitations. One is the vector, which must be accepted by the host cell and also be expressed in the host's genetic processes. The most common vector currently used is called the Ti plasmid vector, which is the disease-producing agent of the bacterium *Agrobac-terium tumefaciens*. This bacterium causes crown gall disease, that is, plant tumors. When used as a vector, the tumor-causing part of the DNA is re-moved and replaced by the genes that the researcher wishes to add to the host cell. The main problem is that this vector only works with plants suscep-tible to crown gall disease. For example, it does not work with such major agricultural crops as corn, rice, and wheat. Viruses could be used as vectors, thus eventually allowing recombinant DNA success with other crops. Much more research is needed, however, with viral vectors.

A second major problem is the regeneration of the host organism from a genetically altered cell. The problem does not occur with single-celled orga-nisms, since regeneration is not needed. Hence we have seen successes with vaccine production using genetically altered bacteria. However, regeneration of a cell into a plant poses considerable difficulty, and into an animal impos-sible difficulties with our present knowledge. Some plants have been regen-erated from cells; these include alfalfa, carrot, petunia, potato, tobacco, and tomato. Others, such as corn, soybeans, and wheat, have not been success-fully regenerated.

Some interesting results have been obtained when protoplasts (plant cells with cell wall removed) of sexually incompatible species were fused. This technique, called protoplast fusion, has given rise to somatic hybrid plants, thus offering the promise of new plants not possible through conventional plant breeding. To date this process works only with closely related species in the same family that are sexually incompatible, but not with unrelated species. At this point protoplast fusion offers no control over what genetic infor-mation is lost or retained, unlike the recombinant DNA method. Even though some novel plants have resulted, such as a cross between tomato and po-tato, these plants need additional breeding and selection by conventional means to produce a plant of use in agriculture. Reviews of protoplast fusion, various resulting somatic hybrids, their genetics and problems have been presented by Shepard *et al.* (1983) and Evans (1983).

Tissue cultures of plants have been moderately successful. Such cultures are useful for the propagation of virus-free plants and the propagation of plants that are either difficult or slow to propagate by conventional methods. Plants processed by commercial tissue culture include asparagus, citrus fruits, orchids, pineapples, and strawberries. However, the difficulty of re-generating whole plants from tissue culture has slowed the use of tissue culture methods with crops such as corn and wheat. Information on the techniques is available in Conger (1981) and Evans *et al. (1983).*

Of great potential value is the mutagenesis of tissue cultures and selection for mutants showing disease resistance, greater yields, or other desirable qualities. In comparison with the selection of mutants by conventional meth-ods, the tissue culture approach offers savings in space, time, labor, and money. No agriculturally useful mutants have been produced yet. The prob-lems with this approach are discussed by Chaleff (1983).

**Tissue cultures often undergo mutation without exposure to mutagens.**

These spontaneous mutations can involve one or many genes, and some are stable in the progeny of the regenerated plants (Evans and Sharp 1983; NRC 1984). Some stable observed traits for wheat include variations in height, number of side shoots, color, and seed storage proteins. This area may offer some promising developments in the future.

What is the likelihood of successful application of biotechnology to agri-culture? The United States is rated as leading the world in the biotechnology race and definite commercial results are expected in agricultural areas (OTA 1984). Vaccines and growth promoters for animal agriculture are predicted to have relatively early commercial success, second only to pharmaceuticals. Success in plant agriculture is not rated among the top three, but is said to hold great promise.

Frozen semen from prize bulls has been used to produce superior dairy cattle.

Today, about 70% of dairy cattle are produced by artificial insemina-tion, but only about 3% of beef cattle. Increased production of farm animals by controlled breeding is even more likely with the advent and recent com-mercialization of frozen embryo transplant technology.

In this method, a cow superior in some traits (e.g., disease resistance or meat quality), is induced with a hormone to produce several times the nor-mal number of eggs. These eggs are fertilized with semen from a superior bull. After 1 week of development, the embryos are removed, stored if neces-sary, and placed in surrogate mothers. In this way production time for a genetically superior cow is greatly reduced (Battelle Memorial Institute 1983). One cow could easily donate enough embryos to produce 100 superior animals in a few years, instead of the equivalent production time under normal circumstances of 100 plus years.

Further multiplication is possible. The embryo can be surgically split, and then developed into identical twins. Another possibility is the fusion of two embryos, such as from two breeds, and the production of a hybrid animal. Such procedures promise rapid upgrading and genetic improvement of da-iry and beef cattle in the near future. Further information on a commercial venture is available (King 1982).

The present limitation is cost; embryo transfers cost about $2000. As costs de-crease, or the values of breeding animals rise, dairy cattle will likely be the first farm animals to be bred mostly by embryo transfer. Beef cattle would follow. Eventually, sheep and swine may be bred by embryo transfer.

In 1981 the amount of energy derived from biomass, including wastes, for use in the United States was 2.7 quad and is projected to reach 3.5 quad by 1985. The 1981 level was about 3.5% of total U.S. energy consumption. This placed biomass ahead of geothermal, solar, and wind energy combined in terms of energy contribution (Haggin and Krieger 1983). An excellent over-view of energy from biomass is available (OTA 1980).

While the energy derived from biomass is impressive, some disquieting questions arise. Is this good or bad for sustainable agriculture and food production? Presently agriculture and forestry utilize about three-fourths of the land in the United States. The amount of surplus land, considering urban and industrial uses, left for increased biomass production appears limited. This suggests diversion of existing land to bio-mass energy at the expense of food, fiber, or wood products. Wastes or crop residues diverted for biomass may increase erosion and reduce soil productivity, since most crop residues and manures now are returned to the soil. One exception is manure pro-duced on feedlots. Other changes might accompany increased biomass production: altered or reduced wildlife habitats, increased costs for land and food, and a rise in biomass-related injuries (Pimentel et *al. 1984).*

Other problems arise. The diversion of food resources is not necessarily efficient. For example, the amount of corn needed to produce ethanol to fuel a car is nine times that needed to feed one individual. A barrel of ethanol produced in Brazil from sugar cane costs twice as much as a barrel of oil (pimentel 1982). A little over one-half of the U.S. corn crop would be needed to supply current U.S. gasoline needs with gasohol (Doering III and Peart 1981).

The future use of biomass to produce energy must be considered care-fully, so that it does not conflict with the goal of sustainable agriculture. One possible benefit from biomass and wastes for sustainable agriculture is on-site energy production with surplus materials left after food and soil needs are satisfied. We will now examine these possibilities. Extensive information on the production of liquid and gaseous fuels from biomass is available (Wise 1981, 1983a, 1983b).

The production of ethanol on the farm through fermentation is possible on a small scale (SERI 1980b, 1981b; Miller 1981). Starting materials can be cereal grains, sugar cane, molasses, sugar beets, wood and wood by-products, cheese whey, potatoes, rice, crop and fruit product residues.

Ethanol produced on the farm is a wet alcohol, since it contains small amounts of water. Production of dry alcohol, which is required in gasohol, requires processing capabilities not likely to exist on farms. However, wet alcohol can be used as a fuel in modified internal combustion engines or by injection into the airstream of a diesel turbocharger as a supplement with diesel fuel.

The use of small amounts of cereal grains, especially if spoiled or of inferior quality, might not be as bad for food production as one might think at first glance. The cereal grain byproduct left after fermentation still retains most of the original nutrients, except for carbohydrates, and could be used in animal feeds and possibly in human foods if the original grain was of high quality.

Crop residues can also be fermented to produce ethanol. The danger here is that most are presently returned to the land for maintenance of soil pro-ductivity. If any are used, they should be surplus; if not, some other form of organic matter should be returned to the soil in their place.

Another alcohol might possibly be produced on the farm: methanol. The process is termed pyrolysis, the heating of biomass with little or no air. Gases and liquids are produced, leaving a charcoal residue. The gasification pro-cess can be modified to maximize methanol production. Biomass used in this process can be rice hulls, nut shells, wheat straw, and wood wastes. Zerbe (1981) presents a discussion of pyrolysis of biomass.

Methanol can be mixed with gasoline to make gasohol. Some engine modifications are required to burn methanol-derived gasohol, since it is more caustic than ethanol or conventional gasohol. Another drawback is legal restrictions; the Environmental Protection Agency will not allow more than 0.3% by volume of methanol because of emission problems (Haggin and Krieger 1983). Unless this changes, the production of ethanol on the farm is more likely, even considering that the starting biomass for methanol is less deleterious for food production and soil maintenance. Ethanol is also more likely to be purchased by processors for processing into gasohol. The farmer could supply energy indirectly, that is, the income could offset energy costs.

**Biogas Production**

The pyrolysis process just described can be modified in the gasification process to maximize gas production. The gaseous mixture, termed gas-ogens, might be used in the future to run stationary engines for irrigation, various farm vehicles, and perhaps boilers. More research is needed for this purpose, since some problems exist (Zerbe 1981).

Methane can be produced from the anaerobic digestion of biomass, in-cluding wastes, and even regional resources such as peat (Aiken *et aL. 1983).* Information on the production of methane is provided by Stafford *et at.* (1980). Such production might be useful on feedlots, where manure accu-mulates in large amounts. A few feedlots presently produce methane (SERI 1981 b). Methane could be used on farms for heating of buildings and water (Fig. 10.4).

Small-scale methane digestors, suitable for farm needs, are possible. How-ever, the scale limits the supply and the gas quality is different from natural gas. The sale of surplus methane is unlikely, both because of limited supply and quality.

The production of oils from seeds for use as a diesel substitute is an interesting possibility (USDA 1981). More research is needed, however, since the substitute fuel leaves an unacceptable carbon residue in engines. The process (pressing of oil) is relatively simple, and the residue is rich in protein and makes an excellent animal feed. Since a large part of farm equipment is diesel powered, this approach could contribute to energy self-sufficiency. Possible seeds for this purpose could include peanuts, cottonseeds, and sunflowers. Researchers at North Dakota State University are investigating the use of sunflower seed oil. The USDA has also developed a process for converting vegetable oil into diesel fuel, which is presently being engine-tested (USDA 1985).

**Assessment of Alternate Energy**

Two possible scenarios exist in regard to farms and alternate forms of energy. Farmers may use alternate forms of energy to approach or even achieve energy self-sufficiency; other farmers may export both energy and food. These are two separate operations that involve different activities and requirements.

The energy exporter would grow certain crops such as trees, corn, or other grains, on a vast scale. These crops would be destined for conversion to alcohols or, less likely, gaseous fuels and probably used for gasohol process-ing. Corn or grain might possibly be modified through breeding for biomass production. For example, yields of wheat for food are constrained because of milling quality requirements; if quality was not required, as in biomass pro-duction, breeders could reportedly increase yields by 30% (Doering III and Peart 1981).

The self-sufficient farm might utilize two or more forms of alternate energy to supplement or possibly replace purchased energy. For example, a mixed crop-livestock farm might use solar energy to heat facilities, provide hot water, and dry grain; produce ethanol to run the pickup truck; and use wind energy to provide some of their electricity.

Large-scale biomass producers may increase the problems of agriculture through competition with food production, degradation of the land through erosion, and competition for water resources. This need not be so if the energy farmers use practices of sustainable agriculture, farm marginal land, and maximize energy production from waste biomass or forest and tree farms. Another interesting possibility is to crop peatlands, swampy areas, and other wetland areas with cattails. The cattail biomass can be converted to gaseous or solid fuels, or used to produce electricity (Aiken *et* al. 1983). A good discussion on wood as biomass and how biotechnology can greatly increase wood yields is available (Garrett 1981; Farnum *et al.* 1983). Energy farms may never develop unless they become economically feasible.

Farmers attempting to achieve energy self-sufficiency will not all utilize the same approach. The forms of alternate energy chosen will be influenced by many variables: economics of conventional versus alternative forms of ener-gy, reliability of conventional energy, geographical location, availability of feedstocks for alternate energy, salability of surplus energy, level of manage-ment sophistication, legal aspects, and results of present and future research. Because of these factors, predicting what forms will be adopted is difficult at best. However, some generalizations are possible (Doering III and Peart 1981).

Wind and solar energy are likely to increase as suppliers of energy on farms. Some applications exist already, and technology is likely to improve in the future. These improvements, cost reductions as the market expands, and relatively low management requirements will make them attractive options for the farmer. In some instances low-head hydropower will appear attractive, because of its reliability and low management requirements. These systems all are likely to generate surplus electricity for sale to utilities. Farm needs will undoubtedly come first, as the selling price will be substantially lower than the purchase price of energy by the farmer.

Biomass energy production will probably not contribute as large a part of on-farm energy over the next decade as will the preceding forms of energy. Even assuming small-scale operations, some biomass is seasonal, the level of required time and management is higher, the materials are bulky, and han-dling can be troublesome. Of the biomass options, the ones most likely to be adopted are those that produce liquid rather than gaseous fuels, since the former are more storable, salable, and usable in vehicles or farm machinery.

Assuming technological improvement, the most likely biomass candidate is oil seed processing to prepare diesel fuel substitute. The fuel will be burned directly, surplus could be sold, and the residue utilized for animal feed on the farm or sold. Since about half of the liquid fuel used on farms is diesel, the contribution could be substantial.

Ethanol production is probably a second contender, since the tech-nologicallevel to produce it in the anhydrous form for gasohol is beyond the farmer's managerial level. Wet alcohol production, however, is possible. Some farmers may find it worthwhile to make the necessary equipment modifications in order to utilize the wet alcohol as a fuel. More likely, it will be more attractive to sell it to gasohol processors and use the cash to offset energy needs.

One factor that may change the use of waste biomass for preparation of alcohols and gaseous fuels is the use of newer wastes to maintain soil productivity, thus freeing up more crop residues and manures as feedstocks. Many wastes are being investigated.

The trend toward larger horsepower tractors is reflected in the five- to six fold increase in horsepower ratings over the last decade in the United States. Today, U.S. firms are making 650-hp tractors and some for-eign manufacturers have introduced a 1000-hp version (McElroy and Krause 1982).

These tractors and their attachments are more efficient in terms of labor and fuel, and probably more economical over the long run. One example cited by Buckingham (1981) is that of a farmer replacing five tractors and tillage machines with one super-tractor. Fuel consumption was cut in half and one person did the work of six before. With such equipment it is possible to disk 259 ha (640 acres) in one long or two medium-length days.

Such machines are suitable for large farm holdings, such as those in the Great Plains, South, and Pacific Northwest, which grow crops such as corn, cotton, rice, soybeans, and wheat. They are not feasible for use with row crop planters. However, these large tractors can be used with chisel plows, cultiva-tors, disk harrows, and drills, and for laying plastic tile drainage lines. Both conventional and no-till operations can use these large tractors.

Major changes also are occurring in planters, which have increased in width about six fold over the last decade. Today's planters may be 24 rows wide and equipped with fertilizer attachments. These machines are highly efficient on large acreages. Special forms of planters for minimum tillage are also appearing now. These planters can sow seed and apply fertilizer, her-bicide, and insecticide in one pass. Such planters minimize energy needs. Fuel cuts of about 75% are claimed for one minimum tillage planter (Mc-Elroy and Krause 1982).

New machinery innovations include a land imprinter designed for restoration of mismanaged and overgrazed grasslands (USDA 1981). In a trial demonstration, grass yields per acre exceeded controls by 58-fold in 10 months. The machine produces a surface resistant to erosion and runoff, giving the simultaneously prepared and sown seedbed a chance to get established.

Another change is the development of a "knife-in" applicator designed for use with fertilizers and pesticides. This machine applies materials to a depth formerly possible only with a moldboard plow. Such a machine will undoubt-edly be of use in no-till operations.

Trends toward larger size and innovative design are also observed with harvesting and on-farm processing equipment. New combines have in-creased harvesting capacity fivefold over the last 5 years. For example, one new combine boosts corn harvesting capacity to 2000 bu/hr. A new cotton harvester reduces cost by one-sixth and doubles the acreage harvested com-pared with smaller machines (McElroy and Krause 1982).

Some crops are not suitable for mechanical harvesting because of varia-tion in cultivar size. In time these crops are also going to be mechanically harvested with new types of machines. One example is the pepper, which varies from small chili to big bell types. Profitability dictates that one mechan-ical harvester should be able to handle all the possible sizes. Such a mechan-ical harvester for peppers is now a reality (Kovalchuk 1983).

Innovation has also produced a harvester and partial processor for peas. This machine harvests the peas, strips the pods, and separates the peas from the debris in one field operation. Fuel and labor costs are reduced by 10-15% (Becker 1983).

One change likely in the future could affect all field machinery from trac-tors to harvesters-the concept of controlled traffic (Battelle Memorial In-stitute 1983). The same paths would be used during each growing season, and would not change from year to year. Equipment, whether used for tillage planting, fertilization, plant protection, or harvesting, would be confined to these paths. This restriction would prevent compaction of the growing area, which causes losses in crop productivity. Significant yield increases have been noted with controlled-traffic farming, except when water is limited. Such a system could also reduce labor if automated guidance systems were utilized.

Another possibility is the incorporation of on-board computers on tractors and planters. Such devices can maximize engine efficiency and save fuel on tractors, when timeliness is not a prime concern. When timeliness is critical, the computer mode can be altered to have the engine produce the maximal work rate. Sensors will monitor field conditions and equipment operation; the feedback will be used to regulate the engine and transmission, thus assuring the desired fuel economy or work rate. Planters will also utilize computers aided by sensors to get uniform crop stands (Battelle Memorial Institute 1983).

Computer innovations that save water and cut costs are already appearing on irrigation equipment. Some of these are laser directed and computer controlled, and use photovoltaic batteries for power (Kanninen 1983). Other recent innovations include computerized neutron field probes to measure root zone moisture, energy-conserving sprinklers, more efficient valves, and better irrigation tubes (Anon. 1983).

A number of factors makes it unlikely that these innovative computer- dependent and larger, more efficient machines will be adopted by all farmers. Some of these factors include the limited acreage of small family farms, topography problems like slopes and field irregularities, roads or bridges unable to handle these large units, plentiful supplies of farm labor, or insuffi-cient cost justification.

**Management**

The management of agricultural production and processing is undergoing an explosive change. The replacement of family farms by corporate farms has already occurred to a significant extent in some areas of agricultural production, particularly broiler production and cattle feedlots. Hog production is likely to go that way too. Intensive management has made these larger operations more efficient and profitable.

Smaller, as well as large farms, now have a technology that can help them become very efficient in management: the computer. More and more com-puter applications are being designed for various areas of agriculture, such as bedding plant producers and vegetable growers, computer access infor-mation systems, and computer-directed equipment. The promise of this new technology for agriculture has attracted the interest of Congress (Congres-sional Research Service 1982).

**Computer-Assisted Management**

Computers are now at a size and price where they can be of assistance to agricultural managers on various levels, from family farms to corporate en-terprises. Computers in agriculture can do three basic things: They can provide information on the managerial level about the local operation, which will make it more efficient; they can provide information from externill sources to solve problems and improve operations; and they can control certain operations, freeing the individual for other tasks.

Computers already are showing great promise in these applications in the bedding plant industry. If utilized correctly, efficiency and profit can be raised. At this point the technology is geared for greenhouse operations with 25,000-50,000 £12 with sales of $250,000-$500,000. It seems likely that in the future computer systems appropriate for smaller operations will be devel-oped (Ball 1983).

The successful adoption of computers in the bedding plant industry is attributed to the fact that appropriate programs and training are available on the commercial level, designed specifically for growers. Applications include cost accounting, crop/ space planning, and inventory control. Software is available for these uses. With this software, growers can get a handle on crop profitability, available (but unnoticed) space, workload distribution, shipping and order information, budget planning, and accounting activities. Ball (1983) has a good discussion of these aspects.

A program is also available to determine which of the many crops a grower handles are the most profitable (Aimone 1983). Additional computer pro-grams are available for environmental control and bench movement in the greenhouse (McPhail 1983). A software program to control the greenhouse environment completely and in a manner to produce the best crop at the most economical cost is already under development. Chrysanthemum growers may be able to use this program in their greenhouses as early as 1985 (Anon. 1983a).

A number of computer-accessible information networks are available and of value to agriculture. Some of these will be described briefly, but more details and a complete listing are available elsewhere (Congressional Research

Service 1982). About two dozen plus networks are in use, and some 15 are in the experimental stages. The USDA (1985) is now actively involved with computer models and systems that will benefit the farmer in the near future.

One well-known network deals in information retrieval, giving bibliographic citations and even abstracts from agricultural publications. This network is called AGRICOLA (Agricultural Online Access) and is sponsored by the U.S. Department of Agriculture and the National Agricultural Library. I might add that it was quite valuable in providing information sources for this book.

Others deal in problem solving, such as the nationwide OSU (Oklahoma State University)· Farm Management Programs. This network specifically solves problems for farm management. Others deal in the marketing of specific commodities, such as the ECI (Egg Clearinghouse, Inc.) and the EMA (Electronic Marketing Association) for lamb marketing networks.

Another provides the latest marketing information, such as weather condi-tions, market conditions, and analyses and recommendations on future prices (INSTANT UPDATE). AGNET is another similar information network. Eloth can supply information via personal computers and cable lV systems. One even deals with the latest research in agriculture and forestry supported by public funds (CRIS, Current Research Information System). Computer networks can be more specific and regional, such as those utilized to sched-ule irrigation in Wisconsin and California. The former is sponsored by the University of Wisconsin Extension Service, while the latter is available through a private management service (Curwen 1984; Ferguson 1984).

Some of the experimental information networks have much to offer and may be converted to full operational status. Some deal with the provision of technical assistance, how-to information, and education for farmers. Others deal with pesticides, integrated pest management, cattle trading, financial aspects, and computer training.

**Computer Constraints**

Computer use on the farm, especially the smaller units or microproces-sors, will undoubtedly increase in the future. Some large corporate farms are using computers now, since they can justify them as cost effective. Estimates by the Kellogg Foundation are that by 1990 some 75% of the commercial farms and 90% of the county extension offices will be utilizing computers. Whether this will become reality depends much upon the resolving of current problems and issues.

The area of computer technology for agriculture is not without problems and issues of controversial nature, as was indicated in depth by the Congres-sional Research Service (1982). It is beyond the scope of this discussion to explore all of these, and only a few will be mentioned.

One problem is perhaps best called "computer anxiety," fear of computers exhibited by some people. The best way to resolve this is through education and simplification of programming. Those presently in farming are reached best through workshops held by county extension offices. Future farmers should be required to be "computer literate" when they graduate from land grant universities. This direction is an imperative if the potential of computers in agriculture is to be realized. Essentially, a generation of farm managers having computer literacy is needed before the full potential of computers can be realized in agriculture.

In some cases private enterprise may provide the training. Since this is a profit venture, the training is usually attached to the purchase of a computer. One such package, which embodies a high degree of integrity, is that pro-vided by Summers un Technical Service (Ball 1983). Besides training, a fol-low-up, toll-free telephone help line is available, as problems develop during later use. This package is for greenhouse management. Because such private operations seek profit, the small farmer or low-cost communication user might not be a suitable market.

Another disturbing aspect is the present confusion about information technology. A bewildering array of computers exists, but which ones are best for the farmer? Where is software designed for the farmer available? Some soft-ware exists for the farmer, but much more development is needed. What information networks for agriculture are available? Who can help the farmer? Who will point out the advantages and cost effectiveness? Who will reach the small farmer with information on this new technology? Who will make the instrumentation and information accessibility available to smaller farmers, including some subsidies if necessary? Will copyright cause problems?

Answers to a few of these questions are now becoming available. For example, at no cost farmers can find answers to three questions. Can my operation benefit from computerization? What microcomputer will suit my needs best? What programs will help me to improve management, record keeping, and bookkeeping? These answers are provided through Apple, IBM, and Radio Shack dealers. The farmer enters information and the AgDisk Demonstration and Evaluator Program gives the answers (Anon. 1984).

The other questions will be answered as the federal and state governments and private enterprise seek out and finalize their roles. Undoubtedly there will be some sharing, such as in information services. Perhaps standardization of equipment will occur. Hopefully, accessibility to computer technology will be available for even small farming units. A big plus would be the creation of a central computer facility, which would serve as a library, clearing house, and distribution center.

Only time and awareness will resolve these issues. As this new technology emerges, Congress is examining its role in agriculture (Congressional Re-search Service 1983). Information technology will play some role in making agriculture more sustainable through improvement of management efficiency.

## REFERENCES

Abel, F. and Walker,B. 1981. Tapping geothermal energy, heat from within the earth. *In* Cutting Energy Costs (The 1980 Yearbook of Agriculture). J. Hayes (Editor). U.S. Dept. Agric., Washington, DC

Aiken, R G., Heichel, G. H., Farnham, R S., Pratt, D. C, Schertz, C E. and Schuler, R T. 1983. Analysis of energy inputs for peat and peatland biomass development. Minn. Agric. Exp. Sm. Tech. Bull. AD- *T8-2209.*

Aimone, T. 1983. Ranking crop profitability. Grower Talks 47 (1) 52-55. ANON. 1983. Items on irrigation. Amer. Fruit Grower *103* (4) 24.

Anon, 1983a. Computerized crops. Amer. Veg. Grower 31 (5) 42-43. Anon, 1984. Free computer help. American Fruit Grower *104* (3) 69.

Argonne National Labouratory, 1983. Argonne Solar Pond Gets State Funding.

News from Argonne National Laboratory, Office of Public Affairs. No. 83-57 (August 2).

BALL, V. 1983. Computers for growers are here. Grower Talks 47 (1) 20-27.

Barton, K. A. and Brill,W. J. 1983. Prospects in plant genetic engineering Science 219, 671-676.

Battelle Memorial Insitute, 1983. Agriculture 2000: A Look at the Future. Battelle Press, Columbus, OH.

Becker, R F. 1983. Mechanical pod strippers. Amer. Veg. Grower 31 (6) 14 .

Bledsoe, B. L. and Henry, Z. A. 1981. Solar drying of large round bales of high moisture hay. *In* Agricultural Energy, Vol. 1. ASAE Publ. 3-81. Amer. Soc. Agric. Eng., St. Joseph, MI.

Borlaug, N. E. 1983. Contributions of conventional plant breeding to food production, Science 219, 689-693.

Buckingham, F. 1981. Big new hitch for big rigs. Farm Show 5 (2) 15.

Butler, J. L. and TROEGER, J. M. 1981. Drying peanuts using solar energy stored in a rock bed. *In* Agricultural Energy, Vol. 1. ASAE Publ. 3-81. Amer. Soc. Agric. Eng., St Joseph, MI.

Chaleff, R S. 1983. Isolation of agronomically useful mutants from plant cell cultures.

Science 219, 676-682.

Coleman, R E. 1981. Building a Low-Cost Solar Dryer Incorporating a Concentrating Reflector.

U.S. Citrus and Subtropical Products Laboratory, Agric. Res. Serv., Winter Haven, FL.

Conger, B. V. 1981. Cloning Agricultural Plants Via in Vitro Techniques. CRC Press, Boca Raton, FL.

Congressional Research Service, 1983. Information Technology for Agricultural America. Report printed for the Committee on Agriculture, U.S. House of Representatives. Superintendent of Documents, Washington, DC

Cundiff, J. S. 1981. A renewable resource system for curing tobacco. *In* Agricultural Ener-gy, Vol. 1. ASAE Publ. 3-81. Amer. Soc. Agric. Eng., St. Joseph, MI.

Curwen,D. 1984. Schedule irrigation with WISP. Amer. Veg. Grower 32 (4) 6-7.

Doe 1983. Hydroelectric Plant Construction Cost and Annual Production Expenses-1980, Department of Energy, Washington, DC

Doering, III, O. C and PEART, R M. 1981. How much extra energy can farms produce. *In* Cutting Energy Costs (The 1980 Yearbook of Agriculture). J. Hayes (Editor). U.S. Dept. Agric., Washington, DC

Evans, D. A. 1983. Agricultural applications of plant protoplast fusion. Bio/T echnology 1 (3) 253-261.

Evans, D. A. and SHARP, W. R 1983. Single gene mutations in tomato plants regenerated from tissue culture. Science 221, 949-951.

Evans, D., Sharp, W. R, Ammirato, P. and YAMADA, Y. 1983. Handbook of Plant Cell Culture: Techniques for Propagation and Breeding. Vol. I. Macmillan, New York.

Farnum, P., Timmis, R. and Kulp, J. L. 1983. Biotechnology of forest yield. Science 219, 694-702.

Ferguson, J. 1984. New firm fine-tunes growers' irrigation. Amer. *Veg.* Grower 32 (4) 8. Futures Group, 1982. Impending Agricultural Technologies: Insights from Leading Sci-entists. *Vols.* 1 and 2. The Futures Group, Glastonbury, CT.

Garrett, L. D. 1981. Forests and woodlands-stored energy for our use. *In* Cutting Ener-gy Costs (The 1980 Yearbook of Agriculture). J. Hayes (Editor). U.S. Dept. Agric., Wash-inqton, DC.

Guceri,S.1. 1983. Solar dehydration. *In* Sustainable Food Systems. D. Knorr (Editor). AVI Publishing Co., Westport, CT.

Haggin, J. and KRIEGER, J. H. 1983. Biomass becoming more important in U.S. energy mix. Chern. Eng. News (March 14),28-30.

Heid, JR., W. G. 1981. An Evaluation of the Young Flat-Plate Solar Collector for Multiple Farm Uses. Agric. Econ. Report No. 466. U.S. Dept, of Agric., Washington, DC.

Heid, JR., W. G. and ALDIS, D. F. 1981. Solar-Supplemented Natural Air Drying of Shelled Corn. Tech. Bull. No. 1654. Econ. and Stat. Service, U.S. Dept. Agric., Washington, DC.

Heid, JR., W. G. and TROTTER, W. K. 1982. Progress of Solar Technology and Potential Farm Uses. Agric. Econ. Rept. No. 489. U.S. Dept. Agric., Washington, DC.

Henson, W. W. 1981. Energy for curing burley tobacco. *In* Agricultural Energy, *Vol.* .

Asae Publ. 3-81. Amer. Soc. Agric. Eng., St. Joseph, MI.

Houghton, D. 1980. Solar goes sophisticated. *In* Successful Farming. Meredith Pub. C.?, Des Moines, IA.

Hung,, B. K. and Toksoy, M. 1981. Greenhouse solar system for effective year-round solar energy utilization in agricultural production. *In* Agricultural Energy, *Vol.* 1. ASAE Publ. 3-81. Amer. Soc. Agric. Eng., St. Joseph, MI.

HUD/DOE. 1979. A Survey of Passive Solar Buildings. U.S. Dept. of Housing and Urban Development and U.S. Dept. of Energy, Washington, DC.

Kanninen, E. 1983. Apply water where and when it's needed. Amer. *Veg.* Grower 31 (6) 17-20.

King, W. 1982. Microscopic techniques have a gigantic effect on cattle breeding industry.

New York Times (December 6), A-16.

Klueter, H. H. 1981. Windpower can save your energy dollars. *In* Cutting Energy Costs (The 1980 Yearbook of Agriculture). J. Hayes (Editor). U.S. Dept. of Agric., Washington, DC.

Kovalchuk,.S. 1983. A mechanical pepper harvester. Amer. *Veg.* Grower 31 (6) 8-9, 22. Kranzier, G. A. 1981. Solar hop drying. *In* Agricultural Energy, *Vol.* 1. ASAE Publ. 3-81.

Matson, W. E. 1981. How we can double hydroelectric power. *In* Cutting Energy Costs (The 1980 Yearbook of Agriculture). J. Hayes (Editor). U.S. Dept. Agric., Washington, DC.

McElroy, R. G. and Krause, K. R. 1982. New Technologies to Raise Agricultural Efficien-cies. Agric. Inf. Bull.' No. 453. U.S. Dept. of Agric., Washington, DC.

McPhail, L. 1983. Motion control in Canada. Oglevee computers-masters of environment.

Grower Talks 47 (1) 36-40, 56-59.

Miller, D. L. 1981. The ABC's of making farm alcohol and gas. *In* Cutting Energy Costs (The 1980 Yearbook of Agriculture). J. Hayes (Editor). U.S. Dept. Agric., Washington, DC.

Miller, L. K., Ungg, A. J. and Bulia, JR., L. A. 1983. Bacterial, viral and fungal insec-ticides. Science 219715-721.

Morrison, D. W. and SHOVE, G. C. 1981. Solar drying of large round hay bales. *In* Agricultural Energy, *Vol.* 1. ASAE Publ. 3-81. Amer. Soc. Agric. Eng., St. Joseph, MI.

NASA. 1982. Capturing Energy from the Wind. National Aeronautics and Space Admin., Washington, DC.

NRC. 1984. Genetic Engineering of Plants: Agricultural Research Opportunities and Policy Concerns. National Academy Press, Washington, DC.

OTA. 1980. Energy from Biological Processes. Office of Technology Assessment, Wash-ington, DC.

OTA. 1982. Genetic Technology: A New Frontier. Office of Technology Assessment. West-view Press, Boulder, CO.

OT A. 1984. Commercial Biotechnology, An International Analysis. Office of Technology As-sessment, Washington, DC.

Schilag, J. H. and SHEPPARD, A. P. 1979. Demonstration of Collectors and Instrumenta-tion for Application to Solar Drying of Peanuts, Tobacco and Forage. Final Report of School of Electrical Engineering, Georgia Institute of Technology, Atlanta. Prepared for the Dept. of Energy, Washington, DC.

SERI. 1980a. Wind Energy Information Directory. Solar Energy Research Institute, Washington, DC.

: SERI. 1980b. Fuel from Farms: A Guide to Small-Scale Ethanol Production. Solar Energy, Research Institute, Washington, DC.

ERI. 1981a. A Guidebook to Renewable Energy Technologies. Solar Energy Research Institute, Washington, DC.

SERI. 1981 b. Fermentation Guide for Common Grains: A Step-by-Step Procedure for Small- Scale Ethanol Fuel Production. Solar Energy Research Institute, Washington, DC.

Shepard, J. F., Bindney, D., Barsby, T. and Kenvke, R. 1983. Genetic transfer in plants through inter-specific protoplast fusion. Science 219 683-688.

Spilljian, C. K. 1981. Capturing and storing energy from the sun. *In* Cutting Energy Costs (The 1980 Yearbook of Agriculture). J. Hayes (Editor). U.S. Dept. Agric., Washington, DC.

Stafford, D. A., HAWKES, D. L. and HORTON, H. R. 1980. Methane Production from Waste Organic Matter. CRC Press, Boca Raton, FL.

Thompson, T. L. and PIERCE, R. D. 1977. Where does solar grain drying fit. *In* Proc. 1977 Solar Grain Drying Conference, Univ. of Illinois, Urbana-Champaign.

Uuniversity of illinois, 1979. lllionis Plan No. SP-546 Portable Solar Collector. Dep.

Agric. Eng., Urbana-Champaign, IL 61801. ($1.)

USDA. 1985. Research Progress in 1984, U.S. Dept. of Agric. Washington, DC.

USDA. 1981. Sunflower Oil May be Farmers' Own Source of Diesel Fuel. Land Imprinter Does Job of Restoring Rangelands. U.S. Dept. Agric., S & E Newsmakers (Science and Education), Sept. (4) 17, 20.

Wise, D. L. 1981. Fuel Gas Production from Biomass. Vols. 1 and 2. CRC Press, Boca Raton, FL.

Wise, D. L. 1983a. Liquid Fuel Developments. CRC Press, Boca Raton, FL. WISE, D. L. 1983b. Fuel Gas Systems. CRC Press, Boca Raton, FL.

Zerbe, J. I. 1981. Turning farm wastes into usable energy. *In* Cutting Energy Costs (The 1980 Yearbook of Agriculture). J. Hayes (Editor). U.S. Dept. Agric., Washington, DC.

# 12

# Development of Agriculture in Egypt

Agricultural origins can be traced to the Neolithic Age (around 8,000 B.C.). Plant domestication seems to have occurred independently in various regions possessing favorable environmental conditions and that include a diversity of habitats in a relatively small area, mild climate, abudant water and wildlife. An important center appears to be Greater Mesopotamia (present day Iraq). This is a rich flood plain where wheat, barley, and oats are believed to have originated. The image below shows the Zagros mountains of Iraq where the landscape appears much as it probably was 10,000 years ago.

The story that emerges from countless studies on the process of domestication is that social groups of hunters/gatherers made a slow transition into a more settled lifestyle closely associated with the regular cultivation of plants, especially cereals. However, see one *anthropological perspective* on the relationship between horticulture and agriculture.

Although crops of relatively minor importance have been developed anew since the Middle Ages, all important crop plants and all domesticated animals were brought under human management ('domestication') in prehistoric times. The process of *domestication* was probably a gradual affair, involving the cultivation of specific groups of plants and the selection ('differential reproduction') of superior types. The first humans to domesticate plants were also the first 'plant breeders' (many believe they were probably women). Plant breeding has been defined as 'applied genetics', but effective breeding was obviously practiced before the principles of genetics were elucidated by Gregor Mendel in the 1850's.

We now know that *major food crops* originated in distinct 'Centers of Origin.' These geographic regions (shown below) were first identified by the great Russian breeder, Alexander Vavilov. The concept of these centers reflects the fact that most crop plants grown throughout the world are not native to the areas of greatest production. The centers of origin are important sources of wild plants related to crop plants which can contribute very useful hereditary traits (genes) such as resistance to diseases and pests. In fact, many of the improved varieties of crop plants carry traits recently obtained from wild species relatives and introduced through selective breeding.

The concept of centers of origins for food crops extends also to the various ornamental plants used in Horticulture. However, because of the tremendous diversity of plants used for ornamental purposes, there aren't centers as clearly delineated as for food crops. For landscape plants, one can identify China/Korea/Japan as a major area of origin (e.g. Rhododendrons, Peonies, etc). Many pot plants have come from South Africa (e.g., geraniums) and the Andean region of South America (e.g. petunias, fuchias, etc).

The domestication of plants and animals paralleled the development of villages, towns, and cities which reached a peak in the Civilizations that developed along the Nile river in present-day *Egypt*. The first formal gardens had their beginnings here. to move on to the next section.

## Agriculture and Antiquity

Western Civilization traces its roots to Mesopotamia, one of the first places where plant domestication is believed to have occurred. Following the domestication of plants and animals (and perhaps as a result of it?), numerous settlements developed in this region. Archeological evidence suggests that the first city, Jericho, was built in the area. Another great 'first' that occurred here was the invention of writing. The great civilizations of Sumeria, Babylon, and Assyria developed in this region. The most important crop plants for all of these civilizations were (and still are in the region)

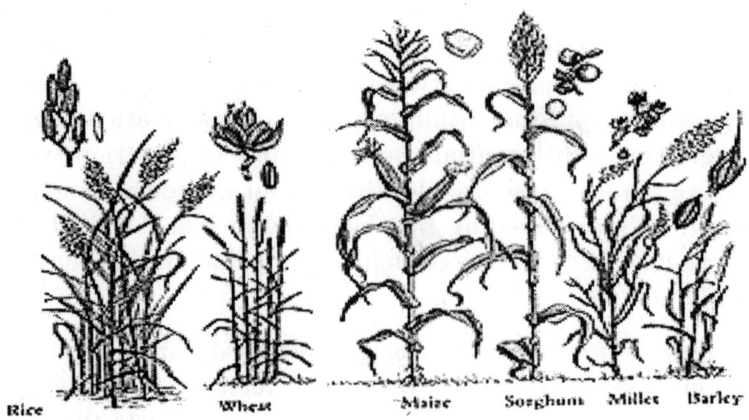

wheat, barley, date, fig, olive, and grape; it is possible that each of them were domesticated from wild plants native to the region. The most advanced civilization of the region, however, arose along the banks of the Nile river...

## Egyptian Civilization

The Egyptians:

Developed various technologies associated with the culinary arts such as: ceramics, the process of baking, wine-making, food storage (pickling, drying, etc)...

Were the first to establish exotic ornamental gardens. That is, they delilberately planned the grounds and placed plants for both ornamental and utilitarian purposes. The first records date to about 2200 BC. The image at right shows a walled garden of an official in Pharaoh's Amenhotep government. The plan is symmetrical and the central, papyrus-fringed pool is shaded by trees. Ornamental plantings were formal, orderly, small, requiring irrigation. Among the elements typical of some gardens were: pools for fish, trees bearing figs, pomegranates, and dates; grapevine-covered trellises; and beds of flowers. Other plants included roses, jasmine and myrtle.

The gardens were primarily those of pharaohs and government officials and those that served as religious or sacred sites.

— developed an 'artificial oasis' that provided privacy (especially for the women of the household) and intimacy cool, leafy, shady gardens were mean to refresh from the intense desert sun a 'human triumph over nature.' This philosophy of mastery over nature, initiated with the Egyptians, found expression in many other cultures as well and is associated with the Formalism school of Landscape Design.

Developed many essential agricultural technologies:

→ drainage (removal of excess water) and irrigation (artificial supply of water) land preparation (refined the hoe and perfected the plow)

Lived in a setting providing the stark contrast of a fertile valley surrounded by desert. Here's a view of the fertile shores of the Nile that give way to the harsh desert in the background as it looks today and as it probably looked at the height of Egyptian Civilization.

Towards the end of Egyptian power, other cultures which would greatly influence Western Civilization were developing in a region now called Israel/Palestine/Lebanon/ Jordan/Syria; this, of course, is the Judeo-Christian Development.

**Industry Overview**

The Indian Agriculture Industry is on the brink of a revolution that will modernize the entire food chain, as the total food production in India is likely to double in the next ten years.

As per recent studies the turnover of the total food market is approximately Rs.250000 crores (US $ 69.4 billion) out of which value-added food products comprise Rs.80000 crores (US $ 22.2 billion). The Government of India has also approved proposals for joint ventures, foreign collaborations, industrial licenses and 100% export oriented units envisaging an investment of Rs.19100 crores (US $ 4.80 billion) out of which foreign investment is over Rs. 9100 crores (US $ 18.2 Billion). The agricultural food industry also assumes significance owing to India's sizable agrarian economy, which accounts for over 35% of GDP and employs around 65 per cent of the population. Both in terms of foreign investment and number of joint- ventures / foreign collaborations, the consumer food segment has the top priority. The other attractive features of the indian agro industry that have the capacity to lure foreigners with promising benefits are the deep sea fishing, aqua culture, milk and milk products, meat and poultry segments.

Excellent export prospects, competitive pricing of agricultural products and standards that are internationally comparable has created trade opportunities in the agro industry. This further has enabled the Indian Agriculture Industry Portal to serve as a means by which every exporter and importer of India and abroad, can fulfill their requirements and avail the benefits of agro related buy sell trade leads and other business opportunities.

This Indian agro industry revolution brings along the opportunities of profitable investment and agriculture-industry-india.com provides you the B2B platform with agro related *catalogs, trade leads, exporters & importers directory* etc. that help you make your way to profit easy.

To lead yourself to the destination of profit through the Indian Agriculture Industry, know maximum about the *EXIM policy, programs & schemes, price policy, seed policy and statistics* at the Indian agro portal and harvest benefits from India, world's second largest producer of food and a country with a billion people. From canned, dairy, processed, frozen food to fisheries, meat, poultry, food grains, alcoholic beverages & soft drinks, the Indian agro industry has dainty areas to choose for business.

# 13

# Geography of Agriculture

Around ten to twelve thousand years ago, human began to domesticate plants and animals for food. Before this first agricultural revolution, people relied on hunting and gathering to obtain food supplies. While there are still groups of hunters and gatherers in the world, most societies have switched to agriculture. The beginnings of agriculture did not just occur in one place but appeared almost simultaneously around the world, possibly through trail and error with different plants and animals or by long term experimentation. Between the first agricultural revolution thousands of years ago and the 17th century, agriculture remained pretty much the same.

In the seventeenth century, a second agricultural revolution took place which increased efficiency of production as well as distribution which allowed more people to move to the cities as the industrial revolution got under way.

The eighteenth century's European colonies became sources of raw agricultural and mineral products for the industrializing nations. Now, many of the countries which were once colonies of Europe, especially those in Central America, are still heavily involved in the same types of agricultural production as they were hundreds of years ago. Farming in the twentieth century has become highly technological in more developed nations with geographical technologies like GIS, GPS, and remote sensing while less developed nations continue with practices which are similar to those developed after the first agricultural revolution, thousands of years ago.

About 45% of the world's population makes their living through agriculture. The proportion of the population involved in agriculture ranges from about 2% in the United States to about 80% in some parts of Asia and Africa. There are two types of agriculture, subsistence and commercial.

There are millions of subsistence farmers in the world, those who produce only enough crops to feed their families. Many subsistence farmers use the slash and burn or swidden agricultural method. Swidden is a technique used by about 150 to 200 million people, and is especially prevalent in Africa, Latin America, and Southeast Asia. A portion of land is cleared and burned to provide at least one and up to three years of good crops for that portion of land. Once the land can no longer be utilized, a new patch of ground is slashed and burnt for another round of crops. Swidden is not a neat or well-organized method of agricultural production by it is effective for farmers who don't know much about irrigation, soil, and fertilization.

The second type of agriculture is commercial agriculture, where the primary purpose is to sell one's product at market. This takes place throughout the world and includes major fruit plantations in Central America as well as huge agribusiness wheat farms in the Midwestern United States.

Geographers commonly identify two major "belts" of crops in the U.S. The wheat belt is identified as crossing the Dakotas, Nebraska, Kansas, and Oklahoma. Corn, which is primarily grown to feed livestock, reaches from southern Minnesota, across Iowa, Illinois, Indiana, and Ohio.

J.H. Von Thunen developed a *model* in 1826 (which wasn't translated into English until 1966) for the agricultural use of land. It has been utilized by geographers since that time. His theory stated that the more perishable and heavier products would be grown closer to urban areas. By looking at the crops grown within metropolitan areas in the U.S., we can see that his theory still holds true. It is very common for perishable vegetables and fruits to be grown within metropolitan areas while less-perishable grain is predominantly produced in non-metropolitan counties.

Agriculture uses about a third of the land on the planet and occupies the lives of about two and a half billion people. It's important to understand where our food comes from.

The geography of Egypt is deeply important in understanding why the Egyptians centered their lives around the Nile. Both before and during the use of canal irrigation in Egypt, the Nile Valley could be separated into two parts, the River Basin or the flat alluvial (or black land soil), and the Red Land or red desert land. The River basin of the Nile was in sharp contrast to the rest of the land of Egypt and was rich with wild life and water fowl, depending on the waxing and waning cycles of the Nile. In contrast, the red desert was a flat dry area which was devoid of most life and water, regardless of any seasonal cycle.

The Nile in it's natural state goes through periods of inundation and relinquishment. The inundation of the Nile-a slightly unpredictable event- was the time of greatest fertility for Egypt. As the banks rose, the water would fill the man-made canals and canal basins and would water the crops for the coming year. However, if the inundation was even twenty inches above or below normal, it could have massive consequences upon the Egyptian agricultural economy. Even with this variability, the Egyptians were able to easily grow tree crops and vegetable gardens in the lower part of the Nile Valley, while at higher elevations, usually near levees, the Nile Valley was sparsely planted.

Agricultural crops were not the mainstay of the ancient Egyptian diet. Rather, the Nile supplied a constant influx of fish which were cultivated year around. In addition to fish, water fowl and cattle were also kept by the Egyptians. Flocks of geese were raised from the earliest times and supplied eggs, meat and fat. However, the domestic fowl didn't make its appearance until Ramesside times, and then in only very isolated places. The Egyptian farmers, in their early experimental phase, also tried to domesticate other animals such as hyenas, gazelles and cranes but gave up after the Old Kingdom. Cattle were also part of the staple diet of the Egyptians, suggesting that grazing land was available for the Egyptians during the times when the Nile receded. However, during the inundation, cattle were brought to the higher levels of the flood plain area and were often fed the grains harvested from the previous year.

The Egyptian diet was by no means limited to tree crops and vegetables, nor was it limited to an animal or fish diet. The Egyptians cultivated barley, emmer wheat, beans, chickpeas, flax, and other types of vegetables. In addition, the cultivation of grains was not entirely for consumption. One of the most prized products of the Nile and of Egyptian agriculture was oil. Oil was customarily used as a payment to workmen employed by the state, and depending on the type, was highly prized. The most common oil (kiki) was obtained from the castor oil plant. Sesame oil from the New Kingdom was also cultivated and was highly prized during the later Hellenistic Period.

The geography of Egypt is deeply important in understanding why the Egyptians centered their lives around the Nile. Both before and during the use of canal irrigation in Egypt, the Nile Valley could be separated into two parts, the River Basin or the flat alluvial (or black land soil), and the Red Land or red desert land. The River basin of the Nile was in sharp contrast to the rest of the land of Egypt and was rich with wild life and water fowl, depending on the waxing and waning cycles of the Nile. In contrast, the red desert was a flat dry area which was devoid of most life and water, regardless of any seasonal cycle.

The Nile in it's natural state goes through periods of inundation and relinquishment. The inundation of the Nile-a slightly unpredictable event- was the time of greatest fertility for Egypt. As the banks rose, the water would fill the man-made canals and canal basins and would water the crops for the coming year. However, if the inundation was even twenty inches above or below normal, it could have massive consequences upon the Egyptian agricultural economy. Even with this variability, the Egyptians were able to easily grow tree crops and vegetable gardens in the lower part of the Nile Valley, while at higher elevations, usually near levees, the Nile Valley was sparsely planted.

Agricultural crops were not the mainstay of the ancient Egyptian diet. Rather, the Nile supplied a constant influx of fish which were cultivated year around. In addition to fish, water fowl and cattle were also kept by the Egyptians. Flocks of geese were raised from the earliest times and supplied eggs, meat and fat. However, the domestic fowl didn't make its appearance until Ramesside times, and then in only very isolated places. The Egyptian farmers, in their early experimental phase, also tried to domesticate other animals such as hyenas, gazelles and cranes but gave up after the Old Kingdom. Cattle were also part of the staple diet of the Egyptians, suggesting that grazing land was available for the Egyptians during the times when the Nile receded. However, during the inundation, cattle were brought to the higher levels of the flood plain area and were often fed the grains harvested from the previous year.

The Egyptian diet was by no means limited to tree crops and vegetables, nor was it limited to an animal or fish diet. The Egyptians cultivated barley, emmer wheat, beans, chickpeas, flax, and other types of vegetables. In addition, the cultivation of grains was not entirely for consumption. One of the most prized products of the Nile and of Egyptian agriculture was oil. Oil was customarily used as a payment to workmen employed by the state, and depending on the type, was highly prized. The most common oil (kiki) was obtained from the castor oil plant. Sesame oil from the New Kingdom was also cultivated and was highly prized during the later Hellenistic Period.

During the 13th century there arose a tendency to commute labour-rents for money payments. This change led to the gradual disappearance of tenants in villeinage—the villeins and cottiers—and the rise on the one hand of the small independent farmer, on the other of the hired labourer. The plague of 1348 marks an epoch in English agriculture. The diminution of the population by one-half led to a scarcity of labour and an increase of wages which deprived the landowner of his narrow margin of profit. To meet this situation, the Statute of Labourers (1351) enacted that no man should refuse to work at the same rate of wages as prevailed before the plague. In addition the landowners attempted to revive the disappearing system of labour-rents. The bitter feelings engendered between employer and employed culminated in the peasants' revolt of 1381.

Meanwhile large numbers of landowners were forced to adopt one of two alternatives. In some cases they ceased to farm their own land and let it out on lease often together with the stock upon it; or else they abandoned arable culture, laid down their demesnes to pasture, enclosed the waste lands and devoted themselves to sheep-farming. In the latter course they were encouraged by the high prices of wool during the 14th century, and by Edward III.'s policy of fostering both the export of wool and the home manufacture of woollen goods. The 15th century, barren of progress in methods of husbandry, was in its early years moderately prosperous. Later on the increasing abandonment of arable husbandry for sheep-farming brought about a less demand for labour, and rural depopulation was accelerated as the peasant was deprived of his grazing-ground by the enclosure of more and more of the waste land.

From the beginning of the reign of Henry VII. to the end of Elizabeth's, a number of statutes were made for the encouragement of tillage, though probably to little purpose. "Where in some towns," says the statute 4th Henry VII. (1488), "two hundred persons were occupied and lived of their lawful labours. now there are occupied two or three herdsmen, and the residue fall into idleness"; therefore it is ordained that houses which within three years have been let for farms, with twenty acres of land lying in tillage or husbandry, shall be upheld, under the penalty of half the profits, to be forfeited to the king or the lord of the fee. Almost half a century afterwards the practice had become still more alarming; and in 1534 a new act was tried, apparently with as little success. "Some have 24,000 sheep, some 20,000 sheep, some 10,000, some 6000, some 4000, and some more and some less"; and yet it is alleged the price of wool had nearly doubled, "sheep being come to a few persons' hands." A penalty was therefore imposed on all who kept above 2000 sheep; and no person was to take in farm more than two tenements of husbandry. By the 39th Elizabeth (1597) arable land made pasture since the 1st Elizabeth shall be again converted into tillage, and what is arable shall not be converted into pasture.

The literature of agriculture, in abeyance since the treatise of Walter of Henley, makes another beginning in the 16th century. The best of the early works is the Book of Husbandry (1st ed. 1523), commonly ascribed to Sir Anthony Fitzherbert, a judge of the Common Pleas in the reign of Henry VIII., but more probably written by his elder brother John. This was followed by the Book of Surveying and Improvements (1523), by the same author. In the former treatise we have a clear and minute description of the rural practices of that period, and from the latter may be learned a good deal of the economy of the feudal system in its decline.

The Book of Husbandry begins with a description of the plough and other implements, after which about a third part of it is occupied with the several operations as they succeed one another throughout the year. Among other passages in this part of the work, the following deserve notice:—

"Somme (ploughs) wyll tourn the sheld bredith at every landsende, and plowe all one way"; the same kind of plough that is now found so useful on hilly grounds. Of wheel-ploughs he observes, that "they be good on even grounde that lyeth lyghte"; and on such lands they are still most commonly employed. Cart-wheels were sometimes bound with iron; of which he greatly approves. On the much agitated question about the employment of horses or oxen in labour, the most important arguments are distinctly stated.

"In some places," he says, "a horse plough is better," and in others an oxen plough, to which, upon the whole, he gives the preference.

Beans and peas seem to have been common crops. He mentions the different kinds of wheat, barley and oats; and after describing the method of harrowing "all maner of cornnes," we find the roller employed. "They used to role their barley grounde after a showr of rayne, to make the grounde even to mowe." Under the article "To falowe," he observes, "the greater clottes (clods) the better wheate, for the clottes kepe the wheat warme all wynter; and at March they will melte and breake and fal in manye small peces, the whiche is a new dongynge and refreshynge of the corne." This is agreeable to the present practice, founded on the very same reasons. "In May, the shepe folde is to be set out"; but Fitzherbert does not much approve of folding, and points out its disadvantages in a very judicious manner. "In the latter end of May and the begynnynge of June, is tyme to wede the corne"; and then we have an accurate description of the different weeds, and the instruments and mode of weeding. Next comes a second ploughing of the fallow; and afterwards, in the latter end of June, the mowing of the meadows begins. Of this operation, and of the forks and rakes and the haymaking there is a very good account. The corn harvest naturally follows: rye and wheat were usually shorn, and barley and oats cut with the scythe. The writer does not approve of the common practice of cutting wheat high and then mowing the stubbles. "In Somersetshire," he says, "they do shere theyr wheat very lowe; and the wheate strawe that they purpose to make thacke of, they do not threshe it, but cut off the ears, and bynd it in sheves, and call it rede, and therewith they thacke theyr houses." He recommends the practice of setting up corn in shocks, with two sheaves to cover eight, instead of ten sheaves as at present—probably owing to the straw being then shorter. The corn was commonly housed; but if there be a want of room, he advises that the ricks be built on a scaffold and not upon the ground. The fallow received a third ploughing in September, and was sown about Michaelmas. "Wheat is moost commonlye sowne under the forowe, that is to say, cast it uppon the falowe, and then plowe it under"; and this branch of his subject is concluded with directions about threshing, winnowing and other kinds of barn-work.

Fitzherbert next proceeds to live stock. "An housbande," he says, "can not well thryue by his corne without he have other cattell, nor by his cattell without corne. And bycause that shepe, in myne opynyon, is the mooste profytablest cattell that any man can haue, therefore I pourpose to speake fyrst of shepe." His remarks on this subject are so accurate that one might imagine they came from a storemaster of the present day.

In some places at present "they neuer seuer their lambes from their dammes"; "and the poore of the peeke (high) countreye, and such other places, where, as they vse to mylke theyr ewes, they vse to wayne theyr lambes at 12 weekes olde, and to mylke their ewes fiue or syxe weekes"; but that, he observes, "is greate hurte to the ewes, and wyll cause them that they wyll not take the ramme at the tvme of the yere for pouertye, but goo barreyne." "In June is tyme to shere shepe; and ere they be shorne, they must be verye well washen, the which shall be to the owner greate profyte in the sale of his wool, and also to the clothe-maker."

The Book of Husbandry begins with a description of the plough and other implements, after which about a third part of it is occupied with the several operations as they succeed one another throughout the year. Among other passages in this part of the work, the following deserve notice:—

"Somme (ploughs) wyll tourn the sheld bredith at every landsende, and plowe all one way"; the same kind of plough that is now found so useful on hilly grounds. Of wheel-ploughs he observes, that "they be good on even grounde that lyeth lyghte"; and on such lands they are still most commonly employed. Cart-wheels were sometimes bound with iron; of which he greatly approves. On the much agitated question about the employment of horses or oxen in labour, the most important arguments are distinctly stated.

"In some places," he says, "a horse plough is better," and in others an oxen plough, to which, upon the whole, he gives the preference.

Beans and peas seem to have been common crops. He mentions the different kinds of wheat, barley and oats; and after describing the method of harrowing "all maner of cornnes," we find the roller employed. "They used to role their barley grounde after a showr of rayne, to make the grounde even to mowe." Under the article "To falowe," he observes, "the greater clottes (clods) the better wheate, for the clottes kepe the wheat warme all wynter; and at March they will melte and breake and fal in manye small peces, the whiche is a new dongynge and refreshynge of the corne." This is agreeable to the present practice, founded on the very same reasons. "In May, the shepe folde is to be set out"; but Fitzherbert does not much approve of folding, and points out its disadvantages in a very judicious manner. "In the latter end of May and the begynnynge of June, is tyme to wede the corne"; and then we have an accurate description of the different weeds, and the instruments and mode of weeding. Next comes a second ploughing of the fallow; and afterwards, in the latter end of June, the mowing of the meadows begins. Of this operation, and of the forks and rakes and the haymaking there is a very good account. The corn harvest naturally follows: rye and wheat were usually shorn, and barley and oats cut with the scythe. The writer does not approve of the common practice of cutting wheat high and then mowing the stubbles. "In Somersetshire," he says, "they do shere theyr wheat very lowe; and the wheate strawe that they purpose to make thacke of, they do not threshe it, but cut off the ears, and bynd it in sheves, and call it rede, and therewith they thacke theyr houses." He recommends the practice of setting up corn in shocks, with two sheaves to cover eight, instead of ten sheaves as at present—probably owing to the straw being then shorter. The corn was commonly housed; but if there be a want of room, he advises that the ricks be built on a scaffold and not upon the ground. The fallow received a third ploughing in September, and was sown about Michaelmas. "Wheat is moost commonlye sowne under the forowe, that is to say, cast it uppon the falowe, and then plowe it under"; and this branch of his subject is concluded with directions about threshing, winnowing and other kinds of barn-work.

Fitzherbert next proceeds to live stock. "An housbande," he says, "can not well thryue by his corne without he have other cattell, nor by his cattell without corne. And bycause that shepe, in myne opynyon, is the mooste profytablest cattell that any man can haue, therefore I pourpose to speake fyrst of shepe." His remarks on this subject are so accurate that one might imagine they came from a storemaster of the present day.

In some places at present "they neuer seuer their lambes from their dammes"; "and the poore of the peeke (high) countreye, and such other places, where, as they vse to mylke theyr ewes, they vse to wayne theyr lambes at 12 weekes olde, and to mylke their ewes fiue or syxe weekes"; but that, he observes, "is greate hurte to the ewes, and wyll cause them that they wyll not take the ramme at the tvme of the yere for pouertye, but goo barreyne." "In June is tyme to shere shepe; and ere they be shorne, they must be verye well washen, the which shall be to the owner greate profyte in the sale of his wool, and also to the clothe-maker."

The Book of Husbandry begins with a description of the plough and other implements, after which about a third part of it is occupied with the several operations as they succeed one another throughout the year. Among other passages in this part of the work, the following deserve notice:—

"Somme (ploughs) wyll tourn the sheld bredith at every landsende, and plowe all one way"; the same kind of plough that is now found so useful on hilly grounds. Of wheel-ploughs he observes, that "they be good on even grounde that lyeth lyghte"; and on such lands they are still most commonly employed. Cart-wheels were sometimes bound with iron; of which he greatly approves. On the much agitated question about the employment of horses or oxen in labour, the most important arguments are distinctly stated.

"In some places," he says, "a horse plough is better," and in others an oxen plough, to which, upon the whole, he gives the preference.

Beans and peas seem to have been common crops. He mentions the different kinds of wheat, barley and oats; and after describing the method of harrowing "all maner of cornnes," we find the roller employed. "They used to role their barley grounde after a showr of rayne, to make the grounde even to mowe." Under the article "To falowe," he observes, "the greater clottes (clods) the better wheate, for the clottes kepe the wheat warme all wynter; and at March they will melte and breake and fal in manye small peces, the whiche is a new dongynge and refreshynge of the corne." This is agreeable to the present practice, founded on the very same reasons. "In May, the shepe folde is to be set out"; but Fitzherbert does not much approve of folding, and points out its disadvantages in a very judicious manner. "In the latter end of May and the begynnynge of June, is tyme to wede the corne"; and then we have an accurate description of the different weeds, and the instruments and mode of weeding. Next comes a second ploughing of the fallow; and afterwards, in the latter end of June, the mowing of the meadows begins. Of this operation, and of the forks and rakes and the haymaking there is a very good account. The corn harvest naturally follows: rye and wheat were usually shorn, and barley and oats cut with the scythe. The writer does not approve of the common practice of cutting wheat high and then mowing the stubbles. "In Somersetshire," he says, "they do shere theyr wheat very lowe; and the wheate strawe that they purpose to make thacke of, they do not threshe it, but cut off the ears, and bynd it in sheves, and call it rede, and therewith they thacke theyr houses." He recommends the practice of setting up corn in shocks, with two sheaves to cover eight, instead of ten sheaves as at present—probably owing to the straw being then shorter. The corn was commonly housed; but if there be a want of room, he advises that the ricks be built on a scaffold and not upon the ground. The fallow received a third ploughing in September, and was sown about Michaelmas. "Wheat is moost commonlye sowne under the forowe, that is to say, cast it uppon the falowe, and then plowe it under"; and this branch of his subject is concluded with directions about threshing, winnowing and other kinds of barn-work.

Fitzherbert next proceeds to live stock. "An housbande," he says, "can not well thryue by his corne without he have other cattell, nor by his cattell without corne. And bycause that shepe, in myne opynyon, is the mooste profytablest cattell that any man can haue, therefore I pourpose to speake fyrst of shepe." His remarks on this subject are so accurate that one might imagine they came from a storemaster of the present day.

In some places at present "they neuer seuer their lambes from their dammes"; "and the poore of the peeke (high) countreye, and such other places, where, as they vse to mylke theyr ewes, they vse to wayne theyr lambes at 12 weekes olde, and to mylke their ewes fiue or syxe weekes"; but that, he observes, "is greate hurte to the ewes, and wyll cause them that they wyll not take the ramme at the tvme of the yere for pouertye, but goo barreyne." "In June is tyme to shere shepe; and ere they be shorne, they must be verye well washen, the which shall be to the owner greate profyte in the sale of his wool, and also to the clothe-maker."

# 14

# Modern Agricultural Practices and Environmental Degradation

In terms of management of resources sustainable agriculture is defined as the management resources for agriculture to satisfy the changing human needs, while maintaining and enhancing the quality of environment and conserving natural resources for future generation.

Ancient man had few needs. There were three basic needs of food, cloth and shelter. Over the years, human population has increased manifold. Along with this, there has been rapid urbanization and industrial and technological development. In this process, our natural resources have been overexploited, and are getting depleted of its vegetation cover and exposed to degradation process, thereby resulting in partial to permanent loss of productivity. Land use is also being changed. Likewise, water table is getting depleted and our environment is getting polluted. No doubt, technological advancement is a dire necessity for achieving self sufficiency in food grains, fodder, fuel wood, and other products. Development, however, must aim at improving the environment without damaging natural resources. Higher standards of living and increased productivity must be achieved without depleting the natural resources and changing the quality of the environment in which we live.

## Ancient Agricultural Practices and their effects on Environment

Early man learnt to lead a settled life only after he had learnt to grow crops at one place. But, even in those times, there were some groups who wandered from one place to another in search for more fertile areas for farming and looking for grazing lands for their domesticated animals. So, these nomadic groups settled for some time in a place, grew crops and grazed their animals. When the fertility of the area was used up and the grass for grazing had been used up, they moved to another place. This led to soil and land degradation in many places. This process continued till the middle ages.

## Modern Agricultural Practices and Environmental Degradation

There are three main factors which have influenced the prevalent modern agricultural practices

(i) Changes due to industrial and agricultural revolutions

(ii) Economic boom in backward countries

(iii) Growth in population.

When industrial revolution occurred in England, is impact was far-reaching and spread to various spheres of life all over the world. The use of certain machines in farming led to a good yield of crops. Many farmers were eager and curious to use new machines and processes to bring changes in agricultural practices. Many experimentations affected the condition of farmlands. Use of new methods were not always useful and beneficial for the soil in the farmed areas. This led to many areas being abandoned or left to turn to wasteland.

Many countries underwent various social and political revolutions. This led to freedom of many countries and certain economic reforms were carried out. This economic boom encouraged industries and agricultural sectors. Provisions for good yielding seeds, loans for farmers to buy seeds, machineries and latest technical facilities for farming became common. Land distribution system changed, as farmers became owners of lands they tilled on and large farm areas came under machine use (tractors, harvesters). Fertilizers made of chemicals were introduced for better yields and chemical pesticides, weedicides came in use. This ruined the condition of the soil. Natural fertility of soil was lost as chemicals wreaked havoc with the soil nutrient cycles.

More people need more food to feed themselves. More food can be grown on larger tracts of land for bumber yields. So, land came in demand. Forests were cut down to make space for farming. Slash and burn farming methods were used in many areas. As deforestation occurred, many areas experienced soil erosion as the trees that held on to the soil were no longer standing. This land also lost its ability to retain water as forests areas are well able to do. But, no forests also meant a fall in the underground watertable.

To grow more crops, more and more chemical fertilizers were used which affected the soil adversely and started land degradation.

## Effects of Claiming land for Agriculture

Farmlands have become scarce as more and more people are taking up these areas as suburbs and small towns that are homes for them. So, for the stupendous task of gaining land for agriculture (to feed more mouths) man has turned to forests and wetlands.

Rainforests are cut down every second in some parts of the world. This cleared area may serve many purposes out of which farming is a major one. Forests which are natural resources and source of many other facilities are dying out due to acid rain or being cut down for gaining land. This is an irreparable loss.

Many rare flora and fauna are found in the rainforests. The vanishing rainforests obviously has led to and is still leading to extinction of many unique plants and animals. Some are in the endangered species list. The role that flora and fauna play in our lives as balance keepers in nature and as sources of beauty, energy and medicines is known to all of us. So, we must stop clearing land for agriculture from rainforests.

Many wetlands that support Unique ecosystems are also being reclaimed for farming. As those areas may have stagnant or estuarine waters mixed in the soil, it is not of good quality for farming. Here, again artificial methods of agriculture and fertilization are needed, leading to further soil pollution.

Mixed Farming

Mixed Farming is defined as the system of farming in which crop production is combined with the rearing of livestock.

Mixed farming provides following advantages to the farmers:

1. It gives more income to the farmers, while reducing the risk of total failure; thus, maximizing the income and minimizing the risks.
2. It provides work to the farmer and family members throughout the year.
3. It ensures more efficient use of land, labour and other resources.
4. The by-products of crops such as straw and fodder are not wasted, as these are used for feeding livestocks on the same farm.
5. Livestocks provide manure in the form of dung, which helps to improve soil-fertility.
6. Food needs of all the family members of a farmer are met with.
7. Mixed farming helps in sustaining agricultural production.

## Mixed Cropping

It is the growing of two or more crops simultaneously intermingled with each other without a definite row pattern with the object of minimizing the risk of crop failure.

- In our country, mixed cropping is an age-old practice.
- In regions of uncertain rainfall, where there is a risk of crop failure due to lack of water, this practice was adopted. The farmers used to grow two or more crops (cereals, pulses and oil crops) together in the field during the early showers of rainfall.
- The basic objective of mixed cropping is to achieve insurance against total crop failure under poor rainfall conditions and thereby minimizing risk and monetary losses.
- Mixed cropping in recent years has been replaced with a new cropping system called intercropping.

Intercropping

Intercropping is growing two or more crops simultaneously in the same field indefinite row pattern with the object of increasing productivity per unit area.

- Intercropping is the practice of growing two or more crops growing two or more crops growing two or more crops simultaneously in the same field in rows.
- Intercropping, there is a greater utilization of the interspaced area, light, nutrients, water and air. As a result, productivity per unit area is increased.
- Intercropping, definite row patterns like 1 : 1, 1 : 2 or 1 :3 are followed, 1:1 row pattern means that there is one row of intercrop. Similarly, in 1 : 2 row pattern, one row of main crop is followed by two rows of intercrops.

Advantages of Intercropping

(1) Productivity is increased.

(2) It economises space and time of cultivating two or more crops.

(3) In case where one of the component crop is a leguminous one, it helps to maintain soil fertility.

(4) There is better utilization of light $CO_2$. rain water and other above ground parameters.

(5) Fertilizers and pesticides can be easily applied to component crop as per the need of the crop.

(6) The component crops can be easily harvested.

(7) Produce of each crop can be marketed and consumed separately.

## Crop Rotation

We have come to know that the practice of growing two or more crops simultaneously in the same field is helpful in overcoming total crop failure under poor rainfall conditions.

However, farmers face some other problems while growing crop in the same field continuously for a number of years. For example.

(i) the soil gets depleted of its nutrients and as a result, the fertility of soil goes down and the productivity of crops is reduced, and

(ii) the weeds, pests and diseases specific to the crop keep on multiplying, bringing about further reduction in crop productivity.

These problems are overcome by growing different types of crops alternately in the same field.

"The practice of growing different crops in the same field alternately, in a pre-planned succession, is called crop rotation."

When different crops are grown on the field on after another, the yields of produce obtained are greater than when the same crops are grown year after year.

Advantages of Crop Rotation

(1) Control of Pests and weeds

(2) Maintenance of sol nitrogen supply

(3) Economy of labour

(4) Protection of soil

(5) Complete use of soil

(6) Orderly farming

(7) Crop-failure risk reduced.

## Impaction Environment

Humans need food as a source of energy. The source of much of the food consumed by man is terrestrial agriculture. This represents the most manipulated of all the non-urban ecosystems, in which the energy and matter pathways are directed almost entirely to man and where he maintains a high level of input of matter and energy to keep the system stable in order to yield his preferred crop. No only is the ecosystem manmade, but the plant and animal made, but the plant and animal components of it have usually been genetically altered by man in the course of their domestication.

Modernization of agriculture leads to several environmental impacts because of the excessive use of pesticides, fertilizers etc.

Many agricultural areas have now large surpluses of plant and animal wastes which when otherwise disposed will cause pollution. Since agriculture is becoming more and more intensive, increasing quantities of fertilizers, pesticides and soil-conditioning agents are used. The wastes formed due to these materials cause severe land pollution problems.

*Fertilizers*: Throughout the world the use of artificial fertilizers on land has increased considerably. Though fertilizers are used to fortify soil, they contaminate the soil to some extent with their impurities.

When the fertilizers are contaminated with other synthetic organic chemicals, the soil water gets polluted.

*Pesticides:* Pesticides help in controlling insect-bonrne diseases and in protection of crops. Among pesticides the most commonly used are the chlorinated hydrocarbons, e.g. DDT, BHC, endrin, aldrin, dieldrin and lindane and organophosphorus compounds, such as parathion and malathion. The remanants of these pesticides may be absorbed by soil particles which contaminate root crops grown in soils. Lindane hasbene reported to taint carrots.

Soil conditioners, fumigants and other chemical agents: In addition to fertilizers and pesticides, soil conditioners and fumigants are used in agriculture. These chemical agents cause alternations in both the agricultural and horticultural lands.

Due to the increasing demand for agricultural practices, forests are destroyed for this purpose. The devastating effects of deforestation in India include soil, water and wind erosions.

Deforestation has a major impact on the productivity of our croplands. This happens in two ways:

(i) soil erosion increases manifold and the soil actually gets washed, leading to an intensified cycle of floods and droughts,

(ii) the shortage of firewood has an obvious impact on the productivity of our croplands.

Destruction of biotic potential of land also leads to desertification Desertification is the process leading to desert formation. Removal of vegetal cover brings about marked changes in the local climates of the areas. Thus, deforestation, overgrazing etc., bring about changes in rainfall temperature, wind velocity, etc., and also leads to soil erosion.

Deforstation has a major impact on the productivity of our croplands. This happens in two ways :

(i) soil erosion increases manifold and the soil actually gets washed, leading to an intensified cycle of floods and droughts.

(ii) The shortage of firewood has an obvious impact on the productivity of our croplands.

Destruction of biotic potential of land also leads to desertification. Desertification is he process leading to desert formation. Removal of vegetal cover brings about marked changes in the local climates of the areas. Thus, deforestation, overgrazing etc., bring about changes in rainfall temperature, wind velocity. Etc., and also leads to soil erosion. Destruction often starts as patchy destruction of productive land.

# Bibliography


Abel, F. and Walker,B. 1981. Tapping geothermal energy, heat from within the earth. *In* Cutting Energy Costs (The 1980 Yearbook of Agriculture). J. Hayes (Editor). U.S. Dept. Agric., Washington, DC

Adkisson,, P. L., Niles, G. A., Walker, J. K., Bird, L. S. and scon, H. B. 1982. Controlling cotion's insect pests: A new system. Science 216, 19-22.

Aiken, R G., Heichel, G. H., Farnham, R S., Pratt, D. C, Schertz, C E. and Schuler, R T. 1983. Analysis of energy inputs for peat and peatland biomass development. Minn. Agric. Exp. Sm. Tech. Bull. AD- *T8-2209.*

Aimone, T. 1983. Ranking crop profitability. Grower Talks 47 (1) 52-55. Anon, 1983. Items on irrigation. Amer. Fruit Grower *103* (4) 24.

Anon, 1983a. Computerized crops. Amer. Veg. Grower 31 (5) 42-43. Anon, 1984. Free computer help. American Fruit Grower *104* (3) 69.

Argonne National Labouratory, 1983. Argonne Solar Pond Gets State Funding. News from Argonne National Laboratory, Office of Public Affairs. No. 83-57 (August 2).

Alberts, E. E., Schuman, G. E. and Burwell, R. E. 1978. Seasonal runoff losses of nitrogen and phosphorus from Missouri *Valley* loess watersheds. J. Environ. Qual. 7 (2), 203-208.

Allen, G. E. 1980. Integrated pest management. Bioscience 30, 655-701.

Andow, D. 1983. Effect of agricultural diversity on insect populations. *In* Environmentally Sound Agriculture. W. Lockeretz (Editor). Praeger Publishers, New York.

Anon. 1983, Connecticut pest management with IPM. Amer. *Veg.* Grower 31 (4) 46. ARAJI, A. A. 1981. The economic impact of investment in integrated pest management. Res. Bull. *115.* Idaho Agric. Exp. Stn.

Batra, S. W. T. 1982. Biological control in agroecosystems. Science 215, 134-139. Bradford, R. R. 1974. Nitrogen and Phosphorus Losses from Agronomy Plots in North Alabama. EPA-660/2-74-033. U.S. Environ. Protection Agency, Washington DC.

Ball, V. 1983. Computers for growers are here. Grower Talks 47 (1) 20-27.

Barton, K. A. and Brill,W. J. 1983. Prospects in plant genetic engineering Science 219, 671-676.

Battelle Memorial Insitute, 1983. Agriculture 2000: A Look at the Future. Battelle Press, Columbus, OH.

Becker, R F. 1983. Mechanical pod strippers. Amer. Veg. Grower 31 (6) 14 .

Bledsoe, B. L. and Henry, Z. A. 1981. Solar drying of large round bales of high moisture hay. *In* Agricultural Energy, Vol. 1. ASAE Publ. 3-81. Amer. Soc. Agric. Eng., St. Joseph, MI.

Borlaug, N. E. 1983. Contributions of conventional plant breeding to food production, Science 219, 689-693.

Buckingham, F. 1981. Big new hitch for big rigs. Farm Show 5 (2) 15.

Butler, J. L. and TROEGER, J. M. 1981. Drying peanuts using solar energy stored in a rock bed. *In* Agricultural Energy, Vol. 1. ASAE Publ. 3-81. Amer. Soc. Agric. Eng., St Joseph, MI.

Brown, G. C. 1982. Microprocessor-based information management system for an inte-grated pest management system. Bull. Entomol. Soc. Amer. 28, 135-137.

Brunner, J. 1981. New tactics for apple IPM. pp. 145-150. *In* Proc. Washington State Horticultural Assoc. 1981.

Caltagirone, L. E. 1981. Landmark examples in classical biological control. Annu. Rev. Entomol. 26, 213-232.

Carnahan, W. B. 1982. Soybean IPM. Extension Rev. 53 (2) 6-7.

Carnahan, W. B. 1982a. Urban IPM: Blooming across the country. Extension Rev. 53 (2), 12-14.

Chaleff, R S. 1983. Isolation of agronomically useful mutants from plant cell cultures. Science 219, 676-682.

Ciesla, W. M. 1982. IPM: New approaches to old problems. Amer. For. 88 (2) 40-44, 51-52.

Coleman, R E. 1981. Building a Low-Cost Solar Dryer Incorporating a Concentrating Reflector. U.S. Citrus and Subtropical Products Laboratory, Agric. Res. Serv., Winter Haven, FL.

Conger, B. V. 1981. Cloning Agricultural Plants Via in Vitro Techniques. CRC Press, Boca Raton, FL.

Congressional Research Service, 1983. Information Technology for Agricultural America. Report printed for the Committee on Agriculture, U.S. House of Representatives. Superintendent of Documents, Washington, DC

Cundiff, J. S. 1981. A renewable resource system for curing tobacco. *In* Agricultural Ener-gy, Vol. 1. ASAE Publ. 3-81. Amer. Soc. Agric. Eng., St. Joseph, MI.

Curwen,D. 1984. Schedule irrigation with WISP. Amer. Veg. Grower 32 (4) 6-7.

Cheng, H. H. 1981. Progress in integrated pest management of tobacco insects in Canada. The Lighter 51 (3) 17-21.

Coleman, E. W. and Ridgway, R. L. 1983. Role of stress tolerance in integrated pest management. *In* Sustainable Food Systems. D. Knorr (Editor). *AVI* Publishing Co., West-port, CT.

Coulson, R. N. 1981. Evolution of concepts of integrated pest management in forests. J. Ga. Entomol. Soc. 16 (Suppl. 1),301-316.

Demichele, D. W. 1975. An evaluation of modelling, systems analysis, and operations reserach in defining agricultural research needs and priorities in pest management. Iowa State J. Res. 49, 597-621.

Doane, C. C. and McManus, M. L. 1981. The Gypsy Moth: Research Toward Integrated Pest Management. Tech. Bull. U.S. Dept. Agric. Forest Service, Washington, DC.

Doe 1983. Hydroelectric Plant Construction Cost and Annual Production Expenses-1980, Department of Energy, Washington, DC

Doering, III, O. C and PEART, R M. 1981. How much extra energy can farms produce. *In* Cutting Energy Costs (The 1980 Yearbook of Agriculture). J. Hayes (Editor). U.S. Dept. Agric., Washington, DC

Duniway, J. M. 1982. Role of biometerology in integrated pest management: Soil, plant, water relations and disease. *In* Biometeorology in Integrated Pest Management. Academic Press, New York.

Eelmer, H. S., Brawner, O. L. and Ewart, W. H. 1981. The citricola scale in San Joaquin IPM programs. Citrograph 66 (6) 137-138, 151.

EPA. 1982. Water Quality Assessment: A Screening Procedure for Toxic and Conventional Pollutants. EPA 600/6-82-004a-c. Center for Environ. Res. Information, U.S. Environ. Pro-tection Agency, Cincinnati, Ohio.

Evans, D. A. 1983. Agricultural applications of plant protoplast fusion. Bio/T echnology 1 (3) 253-261.

Evans, D. A. and SHARP, W. R 1983. Single gene mutations in tomato plants regenerated from tissue culture. Science 221, 949-951.

Evans, D., Sharp, W. R, Ammirato, P. and YAMADA, Y. 1983. Handbook of Plant Cell Culture: Techniques for Propagation and Breeding. Vol. I. Macmillan, New York.

Farnum, P., Timmis, R. and Kulp, J. L. 1983. Biotechnology of forest yield. Science 219, 694-702.

Ferguson, J. 1984. New firm fine-tunes growers' irrigation. Amer. *Veg.* Grower 32 (4) 8. Futures Group, 1982. Impending Agricultural Technologies: Insights from Leading Sci-entists. *Vols.* 1 and 2. The Futures Group, Glastonbury, CT.

Filer, T. H., JR., Solomon, J. D., Cooper, D. T. and Hubbes, M. 1981. Integrated pest management of poplar species. *In* Proc. IX Intern. Congo Plant Protection, 1979. Burgess Pub., Minneapolis.

Fisher, G. C. and Weinzierl, R. 1981. Integrated Pest Management and Its Potential Application in Small Fruits Production. Annu. Rept. Oregon Horticultural Soc., Vol. 72.

Flint, M. L. 1981. Integrated Pest Management for Alfalfa Hay. Statewide IPM Project. IPM

Manual Group. University of California, Berkeley.

Flight, M. L. and Van Den Bosch, R. 1981. Introduction to Integrated Pest Management. Plenum Press, New York.

Fox, J. L. 1983. Soil microbes pose problems for pesticides. Science 22 1, 1029-1031. FRANK, J. R. (Editor) 1981. Integrated pest management: Present and future. Proc. of symposium, sponsored by Weed Control and Pest Management Working Group of the Ameri-can Society for Horticultural Science. HortScience 16, 500-516.

Frecon, J. 1983. IPM saves money. Amer. Fruit Grower *103* (1) 19.

Friend, G. 1983. The potential of an alternative agriculture. *In* Sustainable Food Systems.

D. Knorr (Editor). AVl Publishing Co., Westport, CT.

Garrett, L. D. 1981. Forests and woodlands-stored energy for our use. *In* Cutting Ener-gy Costs (The 1980 Yearbook of Agriculture). J. Hayes (Editor). U.S. Dept. Agric., Wash-inqton, DC.

Gibeault, V. A., Bowen, W. R., OHR, H. D., Thomason, I. J. and Cress, F. 1981.

Guceri,S.1. 1983. Solar dehydration. *In* Sustainable Food Systems. D. Knorr (Editor). AVI Publishing Co., Westport, CT.

Integrated pest management for turf. Calif. Turfgrass Culture 31 (2) 13-15.

Hanson, D. J. 1984. Agricultural uses of ethylene dibromide halted. Chem. & Eng. News (March 5), 13-16.

Hanthorn, M. and Duffy, M. 1983. Returns to Corn Pest Management Practices. Agric. Econ. Rept. No. 501. U.S. Dept. Agric., Washington, DC.

Hartstack, A. W. and WITZ, J. W. 1981. Insect modeling. Agric. Eng. 62 (9) 19-20. Hatfield, J. L. and Thomason, I. J. (Editors). 1982. Biometerology in Integrated Pest Management. Academic Press, New York.

Hodges, R. D. and SCOFIELD, A. M. 1983. Effect of agricultural practices on the health of plants and animals produced: A review. *In* Environmentally Sound Agriculture. W. Lock-eretz (Editor). Praeger Publishers, New York.

Haggin, J. and KRIEGER, J. H. 1983. Biomass becoming more important in U.S. energy mix. Chern. Eng. News (March 14),28-30.

Heid, JR., W. G. 1981. An Evaluation of the Young Flat-Plate Solar Collector for Multiple Farm Uses. Agric. Econ. Report No. 466. U.S. Dept, of Agric., Washington, DC.

Heid, JR., W. G. and ALDIS, D. F. 1981. Solar-Supplemented Natural Air Drying of Shelled Corn. Tech. Bull. No. 1654. Econ. and Stat. Service, U.S. Dept. Agric., Washington, DC.

Heid, JR., W. G. and TROTTER, W. K. 1982. Progress of Solar Technology and Potential Farm Uses. Agric. Econ. Rept. No. 489. U.S. Dept. Agric., Washington, DC.

Henson, W. W. 1981. Energy for curing burley tobacco. *In* Agricultural Energy, *Vol.* . Asae Publ. 3-

81. Amer. Soc. Agric. Eng., St. Joseph, MI.

Houghton, D. 1980. Solar goes sophisticated. *In* Successful Farming. Meredith Pub. C.?, Des Moines, IA.

Hung,, B. K. and Toksoy, M. 1981. Greenhouse solar system for effective year-round solar energy utilization in agricultural production. *In* Agricultural Energy, *Vol.* 1. ASAE Publ. 3-81. Amer. Soc. Agric. Eng., St. Joseph, MI.

HUD/DOE. 1979. A Survey of Passive Solar Buildings. U.S. Dept. of Housing and Urban Development and U.S. Dept. of Energy, Washington, DC.

Khush, G. S. and CHANDHARY, R. C. 1981. Role of Resistant Varieties in Integrated Pest Management of Rice. Extension Bull. No. 162. Food and Fertilizer Technology Center, Taipei City, Taiwan.

Klein, D. T. and KLEIN, R. M. 1985. The growing case against acid rain. Garden 9 (2) 22-27.

Kogan, M. 1981. Dynamics of insect adaptations to soybean: Impact of integrated pest management. Environ. Entomology 10 (3) 363-371.

Kress, L. W. and Miller, J. E. 1983. Impact of ozone on soybean yield. J. Environ. Qual. 72 (2) 276-281.

Kanninen, E. 1983. Apply water where and when it's needed. Amer. *Veg.* Grower 31 (6) 17-20.

King, W. 1982. Microscopic techniques have a gigantic effect on cattle breeding industry. New York Times (December 6), A-16.

Klueter, H. H. 1981. Windpower can save your energy dollars. *In* Cutting Energy Costs (The 1980 Yearbook of Agriculture). J. Hayes (Editor). U.S. Dept. of Agric., Washington, DC.

Kovalchuk,.S. 1983. A mechanical pepper harvester. Amer. *Veg.* Grower 31 (6) 8-9, 22. Kranzier, G. A. 1981. Solar hop drying. *In* Agricultural Energy, *Vol.* 1. ASAE Publ. 3-81.

Kuhr, R. J. 1981. Regional planning and coordination of integrated pest management pro-grams. HortScience 16, 514-515.

Kumar, Arun, *Agriculture Development: Today and Tommow*, Anmol Publications, New Delhi, 2000.

Kydonieus, A. F. and BEROZA, M. 1982. Insect Suppression with Controlled Release Pher-omone Systems. Vols. 1 and 2. CRC Press, Boca Raton, FL.

Lebrun, R. A. 1984. Will *B.b.* beat the beetle? Amer. Veg. Grower 32 (1) 8.

Levenson, H. and Frankie, G. W. 1981. Pest control in the urban environment. Prog. in Resource Management and Environ. Planning 3, 251-272.

Lumsden, R. D., LEWIS, J. A. and Papavizas, G. C. 1983. Effect of organic amendments on soilborne plant diseases and pathogen antagonists. *In* Environmentally Sound Agri-culture. W. Lockeretz (Editor). Praeger Publishers, New York.

Masud, S. M., Lacewell, R. D., Taylor, C. R., BENEDICT, J. H. and L1PPKE, L. A. 1981.

Economic impact of integrated pest management strategies for cotton production in the coastal bend region of Texas. South. J. Agric. Econ. ]3 (2) 47-52.

Mathys, G. 1981. Implementation of integrated pest management on deciduous fruits in Europe. *In* Proc. IX Intern. Congo Plant Protection, 1979. Burgess Pub., Minneapolis.

Matson, W. E. 1981. How we can double hydroelectric power. *In* Cutting Energy Costs (The 1980 Yearbook of Agriculture). J. Hayes (Editor). U.S. Dept. Agric., Washington, DC.

McCarl, B. A. 1981. Economics of integrated pest management: An interpretive review of the literature. Oregon Agric. Exp. Stn. Special Rept.

Metcalf, R. L. and LUCKMANN, W. H. 1982. Introduction to Insect Pest Manatement. 2nd ed. John Wiley & Sons, New York.

McElroy, R. G. and Krause, K. R. 1982. New Technologies to Raise Agricultural Efficien-cies. Agric. Inf. Bull.' No. 453. U.S. Dept. of Agric., Washington, DC.

McPhail, L. 1983. Motion control in Canada. Oglevee computers-masters of environment. Grower Talks 47 (1) 36-40, 56-59.

Miller, D. L. 1981. The ABC's of making farm alcohol and gas. *In* Cutting Energy Costs (The 1980 Yearbook of Agriculture). J. Hayes (Editor). U.S. Dept. Agric., Washington, DC.

Miller, L. K., Ungg, A. J. and Bulia, JR., L. A. 1983. Bacterial, viral and fungal insec-ticides. Science 219715-721.

Morrison, D. W. and SHOVE, G. C. 1981. Solar drying of large round hay bales. *In* Agricultural Energy, *Vol.* 1. ASAE Publ. 3-81. Amer. Soc. Agric. Eng., St. Joseph, MI.

Miller, J. E. 1983. Studies reveal effects of air pollution on crops. Logos] (3) 20-23. MILLER, L. K., LINGG, A. J. and BULLA, L. A. 1983. Bacterial, viral and fungal insecticides. Science 2] 9, 715-721.

Miller, R. L. and NIELSEN, D. G. 1981. Insect and Mite Control on Woody Ornamentals. Ext. Bull. 504. Ohio State Univ.

Musser, W. N., TEW, B. V. and EPPERSON, J. E. 1981. An economic examination of an inte-grated pest management production system with a contrast between E-V and stochas-tic dominance analysis. South. J. Agric. Econ. ]3 (1) 119-124.

Nalewaja, J. D. 1981. Integrated pest management for weed control in wheat. *In* CRC Handbook of Pest Management in Agriculture, Vol. 3. CRC Press, Boca Raton, FL.

NASA. 1982. Capturing Energy from the Wind. National Aeronautics and Space Admin., Washington, DC.

NRC. 1984. Genetic Engineering of Plants: Agricultural Research Opportunities and Policy Concerns. National Academy Press, Washington, DC.

OTA. 1980. Energy from Biological Processes. Office of Technology Assessment, Wash-ington, DC.

OTA. 1982. Genetic Technology: A New Frontier. Office of Technology Assessment. West-view Press,

Boulder, CO.

OT A. 1984. Commercial Biotechnology, An International Analysis. Office of Technology As-sessment, Washington, DC.

Osteen, C. D., BRADLEY, E. B. and Moffitt, L. J. 1981. The Economics of Agricultural Pest Control: An Annotated Bibliography 1960-80. U.S. Dept. Agric. Economics and Statis-tics Service.

OT A. 1979. Pest Management Strategies: Present and Future Pest Management Strategies in the Control of Sorghum and Cotton Pests in Texas. Office of Technology Assessment, Washington, DC.

OT A. 1982. Impacts of Technology on U.S. Cropland and Rangeland Productivity. Office of Technology Assessment, Washington, DC.

OTA. 1982a. Genetic Technology: A New Frontier. Office of Technology Assessrnent. West-view Press, Boulder, CO.

Parr,J. F., Marsh, P. B. and KLA,J. M. 1983. Land Treatment of Hazardous Wastes. Noyes Publications, Park Ridge, NJ.

Payne, C. C. 1982. Insect viruses as control agents. Parasitology 84 (4) 35-77. PIMENTEL, D. and EDWARDS, C. A. 1982. Pesticides and ecosystems. Bioscience 32, 595-600.

Pimentel, D., Glennister, c., FAST, S. and Gallahan, D. 1983. An environmental risk assessment of biological and cultural controls for organic agriculture. *In* Environmentally Sound Agriculture. W. Lockeretz (Editor). Praeger Publishers, New York.

Poe, S. L. 1981. An overview of integrated pest management. *In* HortScience ] 6, 501-506.

Poincelot, R. P. 1980. Plant protection. *In* Horticulture: Principles and Practical Applications. Prentice-Hall, Englewood Cliffs, NJ.

Price, J. F., Englehard, A. W., Overman, Q. W., Yingst, A. J. and IVERSON, M. K. 1981. IPM demonstrations in commercial gypsophola and chrysanthemums. *In* Proc. of the 1980 Meeting, Florida State Horticultural Society.

Putnam, A. R. 1983. Allelopathic chemicals: Nature's herbicides in action. Chem. Eng. News (April 4), 34-45.

Ruesink, W. G. 1976. Status of the systems approach to pest management. Annu. Rev.

Entomo!. 2], 27-44.

Schmidt, R. A. and Wilkinson, R. C. 1981. Prospects for integrated pest management in slash pine ecosystems in Florida. *In* Proc. IX Intern. Congo Plant Protection, 1979. Burgess Pub!., Minneapolis.

Schilag, J. H. and SHEPPARD, A. P. 1979. Demonstration of Collectors and Instrumenta-tion for Application to Solar Drying of Peanuts, Tobacco and Forage. Final Report of School of Electrical Engineering, Georgia Institute of Technology, Atlanta. Prepared for the Dept. of Energy, Washington, DC.

SERI. 1980a. Wind Energy Information Directory. Solar Energy Research Institute, Washington, DC.

SERI. 1980b. Fuel from Farms: A Guide to Small-Scale Ethanol Production. Solar Energy, Research Institute, Washington, DC.

ERI. 1981a. A Guidebook to Renewable Energy Technologies. Solar Energy Research Institute, Washington, DC.

SERI. 1981 b. Fermentation Guide for Common Grains: A Step-by-Step Procedure for Small- Scale Ethanol Fuel Production. Solar Energy Research Institute, Washington, DC.

Shepard, J. F., Bindney, D., Barsby, T. and Kenvke, R. 1983. Genetic transfer in plants through inter-specific protoplast fusion. Science 219 683-688.

Spilljian, C. K. 1981. Capturing and storing energy from the sun. *In* Cutting Energy Costs (The 1980 Yearbook of Agriculture). J. Hayes (Editor). U.S. Dept. Agric., Washington, DC.

Stafford, D. A., HAWKES, D. L. and HORTON, H. R. 1980. Methane Production from Waste Organic Matter. CRC Press, Boca Raton, FL.

Thompson, T. L. and PIERCE, R. D. 1977. Where does solar grain drying fit. *In* Proc. 1977 Solar Grain Drying Conference, Univ. of Illinois, Urbana-Champaign.

Thompson, P. and WHITE, G. B. 1981. An Economic Evaluation of the Potential for Tree Fruit Integrated Pest Management in the Northeast. Dept. of Agricultural Economics, Cor-nell Univ. Agric. Exp. Stn., Ithaca, NY.

Tummla, R. L., Haynes, D. L. and CROFT, B. A. 1976. Modeling for Pest Management:

Concepts, Techniques and Application. Michigan State Univ., E. Lansing.

USDA. 1978. Biological Agents for Pest Control: Status and Prospects. U.S. Dept. Agric., Washington, DC.

USDA. 1980. Basic Principles of Insect Population, Suppression and Management. Agric.

Handbook 512. U.S. Dept. Agric., Washington, DC.

USDA. 1980a. Report and Recommendations on Organic Farming. U.S. Dept. Agric., Washing-ton, DC.

USDA. 1981. America's Soil and Water: Condition and Trends. U.S. Dept. Agric., Washington, DC.

USDA. 1981 a. Cooperative Gypsy Moth Suppression and Regulatory Program Activities. U.S. Dept. Agric., Forest Service, Washington, DC.

USDA. 1981 b. Urban Integrated Pest Management: A Report. U.S. Dept. Agriculture, Science and Education Administration. Extension Committee on Organization and Policy. Univ. of Georgia, Athens.

Uuniversity of illinois, 1979. lllionis Plan No. SP-546 Portable Solar Collector. Dep. Agric. Eng., Urbana-Champaign, IL 61801. ($1.)

USDA. 1985. Research Progress in 1984, U.S. Dept. of Agric. Washington, DC.

USDA. 1981. Sunflower Oil May be Farmers' Own Source of Diesel Fuel. Land Imprinter Does Job of Restoring Rangelands. U.S. Dept. Agric., S & E Newsmakers (Science and Education), Sept. (4) 17, 20.

Van Der Lann, P. A. 1956. The influence of organic manuring on the development of the potato root eelworm, *Heterodera rostochiensis.* Nematology , 113-125.

Weairing, C. H. 1982. Integrated pest management: Progress and prospects with special reference to horticulture. N. Z. J. Exp. Agric. 10 (1) 87-94.

Welch, S. M. 1982. Risk-based design of meteorological networks for integrated pest management. *In* Biometeorology in Integrated Pest Management. Academic Press, New York.

William, R. D. 1981. Complementary interactions betwen weeds, weed control practices, and pests in horticultural systems. HortScience ] 6, 508-513.

Wise, D. L. 1981. Fuel Gas Production from Biomass. Vols. 1 and 2. CRC Press, Boca Raton, FL.

Wise, D. L. 1983a. Liquid Fuel Developments. CRC Press, Boca Raton, FL. WISE, D. L. 1983b. Fuel Gas Systems. CRC Press, Boca Raton, FL.

Zerbe, J. I. 1981. Turning farm wastes into usable energy. *In* Cutting Energy Costs (The 1980 Yearbook of Agriculture). J. Hayes (Editor). U.S. Dept. Agric., Washington, DC.

# Index

***